KEATING ON CONSTRUCTION CONTRACTS

SECOND SUPPLEMENT TO THE NINTH EDITION

KEATING ON CONSTRUCTION CONTRACTS

SECOND SUPPLEMENT TO THE NINTH EDITION

BY

STEPHEN FURST, QC, B.A. (Hons), LL.B. (Hons)
of Middle Temple

THE HON. SIR VIVIAN RAMSEY, M.A.
Formerly one of Her Majesty's Justices, of Middle Temple,
International Judge, Singapore International Commercial Court,
Fellow of the Royal Academy of Engineering, Chartered Engineer,
Fellow of the Institution of Civil Engineers, Honorary Professor,
Department of Civil Engineering, University of Nottingham, Visiting
Professor, Dickson Poon School of Law, King's College, London

With a chapter on Public Procurement

BY

SARAH HANNAFORD, QC, M.A.,
of Middle Temple

a commentary on JCT Forms of Contract

BY

ADRIAN WILLIAMSON, QC, M.A.,
of Middle Temple

and a commentary on the Infrastructure Conditions of Contract

BY

JOHN UFF, C.B.E., QC, B.Sc. (Eng.), Ph.D.,
of Grays Inn
Chartered Engineer, Fellow of the Institution of Civil Engineers, Fellow
of the Chartered Institute of Arbitrators, Fellow Royal Academy of
Engineering, Emeritus Professor of Engineering Law at King's College,
London

SWEET & MAXWELL

Published in 2015 by Thomson Reuters (Professional) UK Limited trading as
Sweet & Maxwell, Friars House, 160 Blackfriars Road, London, SE1 8EZ
(Registered in England & Wales, Company No 1679046.
Registered Office and address for service:
2nd floor, Aldgate House, 33 Aldgate High Street, London EC3N 1DL)

For further information on our products and services, visit
www.sweetandmaxwell.co.uk

Typeset by YHT Ltd, London
Printed and bound by CPI Group (UK) Ltd, Croydon, CR0 4YY

No natural forests were destroyed to make this product; only farmed timber was used and re-planted.

A CIP catalogue record for this book is available from the British Library

ISBN 9780414037328

All rights reserved. Thomson Reuters and the Thomson Reuters Logo are trademarks of Thomson Reuters. Sweet & Maxwell ® is a registered trademark of Thomson Reuters (Professional) UK Limited. Crown Copyright material is reproduced with the permission of the Controller of the HMSO and the Queen's Printer for Scotland.

No part of this publication may be reproduced or transmitted in any form or by any means, or stored in any retrieval system of any nature without prior written permission, except for permitted fair dealing under the Copyright, Designs and Patents Act 1988, or in accordance with the terms of a licence issued by the Copyright Licensing Agency in respect of photocopying and/or reprographic reproduction. Application for permission for other use of copyright material including permission to reproduce extracts in other published works shall be made to the publishers. Full acknowledgment of author, publisher and source must be given.

©
The Estate of Donald Keating and Thomson Reuters
2015

EDITORS' ACKNOWLEDGMENTS

Contributors and researchers:

Kevin Touhey
B.A. (Oxon), M.A. (Bristol)

David Gollancz
B.A. (Sussex)

PUBLISHER'S ACKNOWLEDGMENTS

The commentary on the NEC® form in Ch.21 is reproduced with the kind permission of the NEC® organisation.

The Standard Building Contract with Quantities is reproduced in Ch.19 with the kind permission of the Joint Contracts Tribunal Ltd.

The ICC Form of Contract is reproduced in Ch.20 with the kind permission of the Association for Consultancy and Engineering (ACE).

Extracts from FIDIC are reproduced with the kind permission of the FIDIC organisation.

PREFACE TO THE FIRST SUPPLEMENT TO THE NINTH EDITION

Since we produced the ninth edition in October 2011 there have been a number of significant developments in the general law, in the law relating to construction contracts and in civil litigation procedure. The aim of the supplement is to provide an update dealing with the most important developments.

Whilst most of the significant developments in general law are reported in the official reports and those in construction law are reported in the specialist reports and journals, many cases which illustrate important principles and assist practitioners still remain unreported. These additional cases are therefore included in footnotes whilst the more significant developments generally justify inclusion in the text.

Of particular note in this supplement are the cases in the Supreme Court which continue to define principles of contractual construction: the decision of the Court of Appeal in *P.C. Harrington v Systech* on the contractual basis of payment of adjudicators and the effect of an unenforceable decision; the decision in the Supreme Court in *Benedetti v Swairis* on quantum meruit; the decision in *Walter Lilly v DMW Developments* on claims for extension of time and disruption; and a series of cases further defining aspects of public procurement law.

In civil procedure there have been statutory changes and amendments to the CPR to give effect to the reforms arising out of Lord Justice Jackson's review of the costs of civil litigation. These changes have placed a new focus on the costs of litigation in particular by providing that costs should be proportionate and by introducing costs management as part of the case management process.

This supplement has been produced with the assistance of a small number of members of Keating Chambers and we are very grateful to them for finding time in their busy practices. We also wish to thank those at Sweet & Maxwell who have provided us with help and support: Constance Sutherland who has managed the process with patience; Maureen O'Brien who has carried out essential research; and Lesley Davis who has carefully edited our typescript.

The aim has been to provide an update which states the law up to July 31, 2013.

SF	VR
Keating Chambers,	Royal Court of Justice
15 Essex Street, London	The Rolls Building, London

PREFACE TO THE SECOND SUPPLEMENT TO THE NINTH EDITION

As we observed in the preface to the first supplement in July 2013, there continue to be significant developments in the general law, the law relating to construction contracts and in civil litigation procedure. This second cumulative supplement provides a further update to the ninth edition prior to the next edition.

Of particular note is the Supreme Court decision on representation in *Cramaso v Ogilvie-Grant* and the Court of Appeal decisions on a range of issues: the effect of entire agreement clause in *Mears v Shoreline*, the conflicting decisions in *Aspect v Higgins* and *Walker v Quayside*, now the subject of appeal to the Supreme Court, on the effect of limitation and the availability of restitution in final proceedings arising from temporary adjudication proceedings, on the effect of exclusion clauses on injunctions to prevent breaches of contract in *AB v CD*, on penalty principles to be applied to liquidated damages clauses in *Makdessi v Cavendish Square*, and on the effect on costs of not responding to an offer of mediation in *PGF II v OMFS*.

The Technology and Construction Court has also been tackling familiar problems such as difficulties with contracts founded on estoppel (*Liberty Mercian v Cuddy*), notice provisions (*Obrascon v A-G for Gibraltar*), Defective Premises Act claims (*Rendlesham Estates v Barr*), when adjudication proceedings are commenced (*University of Brighton v Dovehouse*), third party rights to adjudicate (*Hurley Palmer Flatt v Barclays*), and the obligation of due diligence (*Sabic v Punj Lloyd*). Finally, there is the impact of the implementation of the 2014 Directives in the area of public procurement.

This supplement has been produced with the invaluable assistance from Kevin Touhey who undertook the bulk of the research and updating and David Gollancz who prepared a draft of the changes to the public procurement chapter. We also wish to thank those at Sweet & Maxwell who have provided us with help and support: Constance Sutherland who has been in charge of the process, for her encouragement and patience and Lesley Davis who has carefully edited our typescript.

The aim has been to provide an update which states the law up to October 1, 2014.

SF	VR
Keating Chambers, London	London and Singapore

HOW TO USE THIS SUPPLEMENT

This is the Second Supplement to the Ninth Edition of *Keating on Construction Contracts*, and has been compiled according to the structure of the main work.

At the beginning of each chapter of this supplement the mini table of contents from each chapter of the main work has been included. Where a heading in this table of contents has been marked with the symbol ■, this indicates that the material under this heading was updated in the First Supplement. Where headings have been marked with the symbol □, the material under that heading was inserted in this supplement or was updated in the First Supplement but was also updated in this supplement.

Within each chapter, updating information is referenced to the relevant paragraph in the main work.

The law is stated up to October 1, 2014.

CONTENTS

	Page
Editors' Acknowledgments	v
Publisher's Acknowledgments	vii
Preface to the First Supplement to the Ninth Edition	ix
Preface to the Second Supplement to the Ninth Edition	xi
How to use this Supplement	xiii
Table of Cases	xvii
Table of European Cases	xxix
Table of Statutes	xxxi
Table of Statutory Instruments	xxxiii
Table of Civil Procedure Rules	xxxv
Table of European Legislation	xxxvii
Table of References to the JCT Standard Form of Building Contract	xxxix
Chapter 1—The Nature of a Construction Contract	1
Chapter 2—Formation of a Contract	5
Chapter 3—Construction of Contracts	13
Chapter 4—The Right to Payment and Varied Work	23
Chapter 5—Employer's Approval and Architect's Certificates	29
Chapter 6—Excuses for Non-performance	33
Chapter 7—Negligence and Economic Loss	39
Chapter 8—Delay and Disruption Claims	41
Chapter 9—Financial Recovery and Causation	47
Chapter 10—Liquidated Damages and Penalties	55
Chapter 11—Default of the Parties—Various Matters	59
Chapter 12—Various Equitable Doctrines and Remedies	63
Chapter 13—Assignments, Substituted Contracts and Sub-contracts	69
Chapter 14—Architects, Engineers and Surveyors	73
Chapter 15—Public Procurement	79
Chapter 16—Various Legislation	97
Chapter 17—Arbitration	105
Chapter 18—The Housing Grants, Construction and Regeneration Act 1996	113
Chapter 19—Litigation	121
Chapter 20—The JCT Standard Form of Building Contract (2011 edn)	137
Chapter 21—The Infrastructure Conditions of Contract	145

Chapter 22—Engineering and Construction Contract (NEC 3) and
 International Federation of Consulting Engineers Conditions
 (FIDIC)   149

Index   151

TABLE OF CASES

AB v CD. *See* B v D
Ackerman v Ackerman [2011] EWHC 3428 (Ch); [2012] Bus. L.R. D53............ 5–039
Activa DPS Europe Sarl v Pressure Seal Solutions Ltd (t/a Welltec System (UK) [2012]
  EWCA Civ 943; [2012] T.C.L.R. 7; [2012] 3 C.M.L.R. 33; [2012] Eu. L.R. 756.. 9–046
ADS Aerospace Ltd v EMS Global Tracking Ltd [2012] EWHC 2904 (TCC); 145 Con.
  L.R. 29.. 19–008
Adyard Abu Dhabi v SD Marine Services [2011] EWHC 848 (Comm); [2011] B.L.R.
  384; 136 Con. L.R. 190; (2011) 27 Const. L.J. 594......................... 20–100
Ageas (UK) Ltd v Kwik-Fit (GB) Ltd [2014] EWHC 2178 (QB); [2014] Bus. L.R.
  1338.. 9–009
Alexander & Law Ltd v Coveside (21BPR) Ltd [2013] EWHC 3949 (TCC); 152 Con.
  L.R. 163.. 16–040
Alexander v Rayson [1936] 1 K.B. 169; 114 A.L.R. 357 CA...................... 6–051
Alstom Ltd v Yokogawa Australia Pty Ltd (No.7) [2012] SASC 49............... 13–057
Alstom Power Ltd v Somi Impianti Srl [2012] EWHC 2644 (TCC); [2012] B.L.R. 585;
  145 Con. L.R. 17; [2012] C.I.L.L. 3241................................... 11–015
Alstom Transport v Eurostar International Ltd [2012] EWHC 28 (Ch); [2012] 3 All
  E.R. 263; [2012] 2 All E.R. (Comm) 869; [2013] P.T.S.R. 454; 140 Con. L.R. 1;
  [2012] Eu. L.R. 425; (2012) 162 N.L.J. 215....................... 15–009, 15–026
Amey L G Limited v Scottish Minsters [2012] CSOH 181; 2012 G.W.D. 40–775.... 15–018,
                                                                            15–036
Ampurius Nu Homes Holdings Ltd v Telford Homes (Creekside) Ltd; sub nom.
  Telford Homes (Creekside) Ltd v Ampurius Nu Homes Holdings Ltd [2013]
  EWCA Civ 577; [2013] 4 All E.R. 377; [2013] B.L.R. 400; 148 Con. L.R. 1; [2013]
  23 E.G. 76 (C.S.)... 6–077
Andrews v Australia and New Zealand Banking Group Ltd [2013] B.L.R. 111; [2012]
  HCA 30... 10–009
Aspect Contracts (Asbestos) Ltd v Higgins Construction Plc [2013] EWCA Civ 1541;
  [2014] 1 W.L.R. 1220; [2014] Bus. L.R. 367; [2013] 2 C.L.C. 1019; [2014] B.L.R.
  79; 151 Con. L.R. 72; [2014] C.I.L.L. 3449; [2013] 49 E.G. 77 (C.S.)..... 3–040, 16–024
Associated Provincial Picture Houses Ltd v Wednesbury Corp [1948] 1 K.B. 223;
  [1947] 2 All E.R. 680; (1947) 63 T.L.R. 623; (1948) 112 J.P. 55; 45 L.G.R. 635;
  [1948] L.J.R. 190; (1947) 177 L.T. 641; (1948) 92 S.J. 26, CA............. 15–028
Aston Hill Financial Inc v African Minerals Finance Ltd; sub nom. BMA Special
  Opportunity Hub Fund Ltd v African Minerals Finance Ltd [2013] EWCA Civ
  416... 3–002
Atkins Ltd v Secretary of State for Transport [2013] EWHC 139 (TCC); [2013] B.L.R.
  193; 146 Con. L.R. 169; [2013] C.I.L.L. 3337..................... 4–025, 4–052
AXA Insurance Ltd (formerly Winterthur Swiss Insurance Co) v Akther & Darby
  Solicitors [2009] EWCA Civ 1166; [2010] 1 W.L.R. 1662; [2009] 2 C.L.C. 793; 127
  Con. L.R. 50; [2010] Lloyd's Rep. I.R. 393; [2010] Lloyd's Rep. P.N. 187; [2010]
  P.N.L.R. 10; (2009) 106(45) L.S.G. 16; (2009) 159 N.L.J. 1629; [2009] N.P.C.
  129... 9–053, 16–015
B v D; sub nom. AB v CD [2014] EWCA Civ 229; [2014] 3 All E.R. 667; [2014] 2 All
  E.R. (Comm) 242; [2014] C.P. Rep. 27; [2014] B.L.R. 313; 153 Con. L.R. 70;
  [2014] C.I.L.L. 3497.. 3–073A
Baht v Masshouse Developments Ltd [2012] 2 P. & C.R. DG3.................... 8–009

Table of Cases

Balfour Beatty Construction Northern Ltd v Modus Corovest (Blackpool) Ltd [2008] EWHC 3029 (TCC); [2009] C.I.L.L. 2660. 20–234

Bank of Ireland v Philip Pank Partnership [2014] EWHC 284 (TCC); [2014] 2 Costs L.R. 301. 19–077

Barclays Bank Plc v Unicredit Bank AG (formerly Bayerische Hypo– und Vereinsbank AG) [2011] EWHC 3013 (Comm). 19–052

Barr v Biffa Waste Services Ltd [2012] EWCA Civ 312; [2013] Q.B. 455; [2012] 3 W.L.R. 795; [2012] 3 All E.R. 380; [2012] P.T.S.R. 1527; [2012] B.L.R. 275; 141 Con. L.R. 1; [2012] H.L.R. 28; [2012] 2 P. & C.R. 6; [2012] 2 E.G.L.R. 157; [2012] 13 E.G. 90 (C.S.); (2012) 109(14) L.S.G. 21; (2012) 156(12) S.J.L.B. 31. 11–038

BDMS Ltd v Rafael Advanced Defence Systems [2014] EWHC 451 (Comm); [2014] 1 Lloyd's Rep. 576. 6–069

Bell v Peter Browne & Co [1990] 2 Q.B. 495; [1990] 3 W.L.R. 510; [1990] 3 All E.R. 124; (1990) 140 N.L.J. 701 CA (Civ Div). 16–015

Belmont Park Investments Pty v BNY Corporate Trustee Services Ltd [2011] UKSC 38; [2012] 1 A.C. 383; [2011] 3 W.L.R. 521; [2012] 1 All E.R. 505; [2011] Bus. L.R. 1266; [2011] B.C.C. 734; [2012] 1 B.C.L.C. 163; [2011] B.P.I.R. 1223. 16–032

Benedetti v Sawiris [2013] UKSC 50; [2013] 3 W.L.R. 351. .... 4–020, 4–021

Berent v Family Mosaic Housing [2012] EWCA Civ 961; [2012] B.L.R. 488; [2012] C.I.L.L. 3213. 11–044

Berezovsky v Abramovich [2011] EWCA Civ 153; [2011] 1 W.L.R. 2290; [2011] 1 C.L.C. 359; (2011) 108(10) L.S.G. 23. 16–022

Berrisford v Mexfield Housing Co-operative Ltd [2011] UKSC 52; [2012] 1 A.C. 955; [2011] 3 W.L.R. 1091; [2012] 1 All E.R. 1393; [2012] P.T.S.R. 69; [2012] H.L.R. 15; [2012] 1 P. & C.R. 8; [2012] L. & T.R. 7; [2011] 3 E.G.L.R. 115; [2011] 46 E.G. 105 (C.S.); (2011) 155(43) S.J.L.B. 35; [2011] N.P.C. 115; [2012] 1 P. & C.R. DG10. 3–002

Bexhill UK Ltd v Razzaq [2012] EWCA Civ 1376. 13–005

BICC plc v Burndy Corp [1985] Ch. 232; [1985] 2 W.L.R. 132; [1985] 1 All E.R. 417; [1985] R.P.C. 273; (1984) 81 L.S.G. 3011; (1984) 128 S.J. 750, CA. 11–011

Bluewater Energy Services BV v Mercon Steel Structures BV [2014] EWHC 2132 (TCC); 155 Con. L.R. 85. .... 9–043, 19–021

Bocardo SA v Star Energy UK Onshore Ltd; sub nom. Star Energy UK Onshore Ltd v Bocardo SA; Star Energy Weald Basin Ltd v Bocardo SA [2010] UKSC 35; [2011] 1 A.C. 380; [2010] 3 W.L.R. 654; [2010] 3 All E.R. 975; [2011] B.L.R. 13; [2010] 3 E.G.L.R. 145; [2010] R.V.R. 339; [2010] 31 E.G. 63 (C.S.); [2010] N.P.C. 88. 11–046

Boycott v Perrins Guy Williams [2011] EWHC 2969 (Ch); [2012] P.N.L.R. 25. 16–015

BP Exploration Co (Libya) Ltd v Hunt (No.2) [1979] 1 W.L.R. 783; (1979) 123 S.J. 455. 4–021

Bremer Vulkan Schiffbau und Maschinenfabrik v South India Shipping Corp Ltd; Gregg v Raytheon; sub nom. Bremer Vulcan Schiffbau und Maschinenfabrik v South India Shipping Corp [1981] A.C. 909; [1981] 2 W.L.R. 141; [1981] 1 All E.R. 289; [1981] 1 Lloyd's Rep. 253; [1981] Com. L.R. 19; [1981] E.C.C. 151; (1981) 125 S.J. 114 HL. 6–069

Brighton University v Dovehouse Interiors Ltd; sub nom. University of Brighton v Dovehouse Interiors Ltd [2014] EWHC 940 (TCC); [2014] B.L.R. 432; 153 Con. L.R. 147. .... 5–028, 20–036

Brit Inns Ltd (In Liquidation) v BDW Trading Ltd [2012] EWHC 2143 (TCC); 145 Con. L.R.181. 9–013

British Transport Commission v Gourley [1956] A.C. 185; [1956] 2 W.L.R. 41; [1955] 3 All E.R. 796; [1955] 2 Lloyd's Rep. 475; 49 R. & I.T. 11; (1955) 34 A.T.C. 305; [1955] T.R. 303; (1956) 100 S.J. 12 HL. 9–020

BSS Group PLC v Makers (UK) Ltd (t/a Allied Services) [2011] EWCA Civ 809; [2011] T.C.L.R. 7. 3–053

By Development Limited v Covent Garden Market Authority [2012] EWHC 2546 (TCC); 145 Con. L.R. 102. 15–028

Campbell v Daejan Properties Ltd [2012] EWCA Civ 1503; [2013] H.L.R. 6; [2013] 1 P. & C.R. 14; [2013] 1 E.G.L.R. 34; [2013] 6 E.G. 106; [2012] 48 E.G. 62 (C.S.). .. 3–023

Carey Value Added SL v Grupo Urvasco (2010) 132 Con. L.R. 15. .... 11–035, 11–036

TABLE OF CASES

Cavenagh v William Evans Ltd [2012] EWCA Civ 697; [2013] 1 W.L.R. 238; [2012] 5 Costs L.R. 835; [2012] I.C.R. 1231; [2012] I.R.L.R. 679; (2012) 156(23) S.J.L.B. 35. 4–014
Cavendish Square Holdings BV v Makdessi [2012] EWHC 3582 (Comm); [2013] 1 All E.R. (Comm) 787. 11–011
Chandler v Cape Plc [2012] EWCA Civ 525; [2012] 1 W.L.R. 3111; [2012] 3 All E.R. 640; [2012] I.C.R. 1293; [2012] P.I.Q.R. P17. 7–004
Chester Grosvenor Hotel Co v Alfred McAlpine Management (1991) 56 B.L.R. 115. 3–078
Chigwell (Shepherds Bush) Ltd v ASRA Greater London Housing Association Ltd [2012] EWHC 2746 (QB) 15–036
Cine Bes Filmcheck ve Yapimcilik AS v United International Pictures. *See* United International Pictures v Cine Bes Filmcilik ve Yapimcilik AS
CIP Property (AIPT) Ltd v Transport for London [2012] B.L.R. 202 (Ch.). 12–020
Claymore Services Ltd v Nautilus Properties Ltd [2007] EWHC 805 (TCC); [2007] B.L.R. 452. 4–021
Cleightonhills v Bembridge Marine Ltd [2012] EWHC 3449 (TCC); [2013] C.I.L.L. 3289. 7–009
Cleveland Bridge UK Ltd v Severfield–Rowan Structures Ltd [2012] EWHC 3652 (TCC). 8–062
Clinical Solutions International Ltd v NHS 24 [2012] CSOH 10; [2012] Eu. L.R. 398; 2012 G.W.D. 6–116. 15–036
Clinton (t/a Oriel Training Services) v Department of Employment and Learning [2012] NICA 48. .... 15–022, 15–028
Co-operative Group Ltd v Birse Developments Ltd (In Liquidation) [2014] EWHC 530 (TCC); [2014] B.L.R. 359; 153 Con. L.R. 103; [2014] P.N.L.R. 21. ...... 13–019, 16–015
Cometson v Merthyr Tydfil CBC [2012] EWHC 3446 (Ch); [2012] 50 E.G. 101 (C.S.). 3–041
Compagnie Financiere du Pacifique v Peruvian Guano Co (1882) 11 QBD 55, CA. . 19–049
Compass Group UK and Ireland Ltd (t/a Medirest) v Mid Essex Hospital Services NHS Trust; sub nom. Mid Essex Hospital Services NHS Trust v Compass Group UK and Ireland Ltd (t/a Medirest) [2013] EWCA Civ 200; [2013] B.L.R. 265; [2013] C.I.L.L. 3342. 3–043
Connell v Mutch (t/a Southey Building Services) [2012] EWCA Civ 1589; [2013] C.P. Rep. 11; [2013] B.L.R. 82; [2013] T.C.L.R. 1. 19–090
Corelogic Ltd v Bristol CC [2013] EWHC 2088 (TCC). 15–031, 15–032
Covanta Energy Ltd v Merseyside Waste Disposal Authority [2013] EWHC 2922 (TCC); 151 Con. L.R. 146. .... 15–027, 15–030
Covanta Energy Ltd v Merseyside Waste Disposal Authority Unreported. 15–030
Cramaso LLP v Viscount Reidhaven's Trustees; sub nom. Cramaso LLP v Ogilvie-Grant; Cramaso LLP v Earl of Seafield [2014] UKSC 9; [2014] A.C. 1093; [2014] 2 W.L.R. 317; [2014] 2 All E.R. 270; [2014] 1 All E.R. (Comm) 830; 2014 S.C. (U.K.S.C.) 121; 2014 S.L.T. 521; 2014 S.C.L.R. 484; (2014) 158(7) S.J.L.B. 37. .... 6–019, 7–027
Crema v Cenkos Securities Plc [2010] EWCA Civ 1444; [2011] 1 W.L.R. 2066; [2011] 2 All E.R. (Comm) 676; [2011] Bus. L.R. 943; [2010] 2 C.L.C. 963; [2011] C.I.L.L. 2980. 3–013
Crossco No.4 Unlimited v Jolan Ltd [2011] N.P.C 38 (Ch). 12–004
Cukorova Finance International Ltd v Alfa Telecom Turkey Ltd [2013] UKPC 2. . . 11–011
Dany Lions Ltd v Bristol Cars Ltd [2014] EWHC 817 (QB); [2014] 2 All E.R. (Comm) 403; [2014] Bus. L.R. D11. 2–021
Denton v TH White Ltd; Utilise TDS Ltd v Cranstoun Davies; Decadent Vapours Ltd v Bevan [2014] EWCA Civ 906; [2014] 1 W.L.R. 3926; [2014] C.P. Rep. 40; [2014] B.L.R. 547; 154 Con. L.R. 1; [2014] 4 Costs L.R. 752; (2014) 164(7614) N.L.J. 17. .... 19–055, 19–077
Dillard v F&C Commercial Property Holdings Ltd [2014] EWHC 1219 (QB); [2014] C.I.L.L. 3517. 16–059
Direct Travel Insurance v McGeown. *See* McGeown v Direct Travel Insurance
Doosan Babcock Ltd v Comercializadora de Equipos y Materiales Mabe Lda (formerly Mabe Chile Lda) [2013] EWHC 3201 (TCC); [2014] B.L.R. 33. 11–036

TABLE OF CASES

Doosan Babcock Ltd v Comercializadora de Equipos y Materiales Mabe Lda (formerly Mabe Chile Lda) [2013] EWHC 3010 (TCC); [2014] 1 Lloyd's Rep. 464. . 12–020
Dowland v Architects Registration Board [2013] EWHC 893 (Admin); [2013] B.P.I.R. 566; (2013) 163(7558) N.L.J. 16... 14–007
Downing v Al Tameer Establishment [2002] EWCA Civ 721; [2002] 2 All E.R. (Comm) 545; [2002] C.L.C. 1291; [2002] B.L.R. 323................................. 6–069
DW Moore v Ferrier [1988] 1 W.L.R. 267; [1988] 1 All E.R. 400; (1988) 132 S.J. 227 CA (Civ Div).. 16–015
Easycoach Ltd v Department for Regional Development [2012] NIQB 10.......... 15–016, 15–028, 15–029, 15–032
Eaton v Natural England [2012] EWHC 2401 (Admin); [2013] 1 C.M.L.R. 10; [2013] Env. L.R. 9... 16–028
Elekta Ltd v Common Services Agency [2011] CSOH 107; 2011 S.L.T. 815; 2011 G.W.D. 22–507... 15–036
Elvanite Full Circle Ltd v AMEC Earth & Environmental (UK) Ltd [2013] EWHC 1191 (TCC); 148 Con. L.R. 127...................................... 16–014, 19–077
Farstad Supply AS v Enviroco Ltd [2010] UKSC 18; [2010] Bus. L.R. 1087; [2010] 2 Lloyd's Rep. 387; 2010 S.C. (U.K.S.C.) 87; 2010 S.L.T. 994; 2010 S.C.L.R. 379; [2010] 1 C.L.C. 692.. 3–073
Fenice Investments Inc v Jerram Falkus [2010] 128 Con. L.R. 124................. 20–029
FG Wilson (Engineering) Ltd v John Holt & Co (Liverpool) Ltd; sub nom. Caterpillar (NI) Ltd (formerly FG Wilson (Engineering) Ltd) v John Holt & Co (Liverpool) Ltd [2013] EWCA Civ 1232; [2014] 1 W.L.R. 2365; [2014] 1 All E.R. 785; [2014] 1 All E.R. (Comm) 393; [2014] 1 Lloyd's Rep. 180; [2013] 2 C.L.C. 501; [2014] B.L.R. 103; [2014] B.P.I.R. 1104;................................. 3–078, 11–014, 19–064
Financial Services Authority v Sinaloa Gold Plc [2013] 2 W.L.R. 678 (SC)......... 12–021
Fitzroy House Epworth Street (No.1) Ltd v Financial Times Ltd [2006] EWCA Civ 329; [2006] 1 W.L.R. 2207; [2006] 2 All E.R. 776; [2006] 2 P. & C.R. 21; [2006] L. & T.R. 20; [2006] 2 E.G.L.R. 13; [2006] 19 E.G. 174; [2006] 14 E.G. 175 (C.S.); [2006] N.P.C. 40... 3–022
Forbes v Git [1922] 1 A.C. 256, PC (Can)....................................... 3–034
Forster v Outred & Co [1982] 1 W.L.R. 86; [1982] 2 All E.R. 753; (1981) 125 S.J. 309 CA (Civ Div).. 16–015
Fortress Valley Recovery Fund I LLC v Blue Skye Special Opportunities Fund LP [2013] EWCA 367... 13–022
Foxholes Nursing Home Ltd v Accora Ltd [2013] EWHC 3712 (Ch).............. 4–007
Freetown Ltd v Assethold Ltd [2012] EWCA Civ 1657; [2013] 1 W.L.R. 701; [2013] 2 All E.R. 323; [2013] C.P. Rep. 16; [2013] 1 E.G.L.R. 57; [2013] 11 E.G. 82; [2013] R.V.R. 150; [2013] 1 E.G. 49 (C.S.); [2013] 1 P. & C.R. DG13................ 16–063
Gallagher v ACC Bank Plc [2012] IESC 35; [2012] PNLR 29..................... 16–015
Geden Operations Ltd v Dry Bulk Handy Holdings Inc (The Bulk Uruguay); Bulk Uruguay, The [2014] EWHC 885 (Comm); [2014] 2 All E.R. (Comm) 196; [2014] 2 Lloyd's Rep. 66; (2014) 164(7601) N.L.J. 20.................................. 6–066
General Tire & Rubber Co v Firestone Tire & Rubber Co Ltd (No.2) [1975] 1 W.L.R. 819; [1975] 2 All E.R. 173; [1975] F.S.R. 273; [1976] R.P.C. 197; (1975) 119 S.J. 389, HL... 4–021
George Wimpey UK Ltd (formerly Wimpey Homes Holdings Ltd) v VI Construction Ltd (formerly VI Components Ltd); sub nom. George Wimpey UK Ltd (formerly Wimpey Homes Holdings Ltd) v VIC Construction Ltd (formerly Vic Components Ltd) [2005] EWCA Civ 77; [2005] B.L.R. 135; 103 Con. L.R. 67; (2005) 102(9) L.S.G. 28; (2005) 149 S.J.L.B. 182; [2005] 2 P. & C.R. DG5............ 12–013
Giles v Tarry [2012] EWCA Civ 837; [2012] 2 P. & C.R. 15; [2012] 3 E.G.L.R. 5; [2012] 40 E.G. 114; (2012) 156(25) S.J.L.B. 31....................................... 3–022
Glasgow Rent Deposit & Support Scheme v Glasgow CC [2012] CSOH 199; 2013 G.W.D. 2–58... 15–036
Gleeds Retirement Benefits Scheme, Re; sub nom. Briggs v Gleeds [2014] EWHC 1178 (Ch); [2014] 3 W.L.R. 1469; [2014] Pens. L.R. 265; (2014) 158(18) S.J.L.B. 37... 12–002
Golden Eye (International) Ltd v Telefonica UK Ltd [2012] EWCA Civ 1740; [2013] Bus. L.R. 414; [2013] 2 C.M.L.R. 27; [2013] R.P.C. 18......................... 13–020

Greater Vancouver Water District v North American Pipe & Steel Ltd (2012) BCCA
  337 CA (BC).. 3–054
Green v Eadie [2012] Ch. 363; [2012] 2 W.L.R. 510; [2012] P.N.L.R. 9.............. 16–015
Greenwich Millennium Village Ltd v Essex Services Group Plc (formerly Essex Electrical Group Ltd); sub nom. HS Environment Services Ltd v DG Robson Mechanical Services Ltd [2014] EWCA Civ 960; [2014] 1 W.L.R. 3517; 156 Con.
  L.R. 1... 3–065, 9–055, 20–313
Greer v Kettle; sub nom. Parent Trust & Finance Co Ltd, Re [1938] A.C. 156; [1937] 4
  All E.R. 396; (1937) 107 L.J. Ch. 56; (1937) 158 L.T. 433 HL................. 3–061
Hackney Empire Ltd v Aviva Insurance UK Ltd (formerly t/a Norwich Union
  Insurance Ltd) [2012] EWCA Civ 1716; [2013] B.L.R. 57; 146 Con. L.R. 1; [2013]
  1 E.G.L.R. 101; [2013] 2 E.G. 66 (C.S.)....................................... 20–362
Hackney Empire Ltd v Aviva Insurance UK Ltd [2012] EWCA Civ 1716; 146 Con.
  L.R. 1.. 11–026, 11–033
Hall v Van Der Heiden [2010] EWHC 586 (TCC)................................. 20–120
Hammersmatch Properties (Welwyn) Ltd v Saint-Gobain Ceramics & Plastics Ltd
  [2013] EWHC 2227 (TCC); [2013] B.L.R. 554; 149 Con. L.R. 147; [2013] 5 Costs
  L.R. 758; [2013] 3 E.G.L.R. 123; (2013) 163(7571) N.L.J. 20.................. 19–088
Harrison v Shepherd Homes Ltd [2011] EWHC 1811 (TCC); (2011) 27 Const. L.J.
  709... 3–058, 16–002
Hawksford Trustees Jersey Ltd v Stella Global UK Ltd [2012] EWCA Civ 55; [2012] 2
  All E.R. (Comm) 748.. 12–013
Healthcare at Home Limited v Common Services Agency [2011] CSOH 22; 2011
  G.W.D. 8–197... 15–030
Healthcare at Home Ltd v Common Services Agency [2014] UKSC 49; [2014] 4 All
  E.R. 210; [2014] P.T.S.R. 1081; 2014 S.C. (U.K.S.C.) 247; 2014 S.L.T. 769; [2015]
  1 C.M.L.R. 12; 2014 G.W.D. 25-505... 15–016
Henry Brothers (Magherafelt) Ltd v Department of Education for Northern Ireland
  [2011] NICA 59; [2012] B.L.R. 36..................................... 15–016, 15–032
Henry v News Group Newspapers Ltd [2013] EWCA Civ 19; [2013] 2 All E.R. 840;
  [2013] C.P. Rep. 20; [2013] 2 Costs L.R. 334; (2013) 163 N.L.J. 140; (2013) 157(5)
  S.J.L.B. 31.. 19–077
Higginson Securities (Developments) Ltd v Hodson [2012] EWHC 1052 (TCC); [2012]
  B.L.R. 321; [2012] T.C.L.R. 6; 142 Con. L.R. 117; [2012] C.I.L.L. 3177....... 19–005
HIH Casualty & General Insurance Ltd v Chase Manhattan Bank [2003] UKHL 6;
  [2003] 1 All E.R. (Comm) 349; [2003] 2 Lloyd's Rep. 61; [2003] 1 C.L.C. 358;
  [2003] Lloyd's Rep. I.R. 230; (2003) 147 S.J.L.B. 264........................ 3–065
Hillcrest Homes Ltd v Beresford & Curbishley Ltd [2014] EWHC 280 (TCC); 153 Con.
  L.R. 179; [2014] C.I.L.L. 3506... 20–405
Hoad & Taylor v Delves [2012] EWHC 1426 (QB)................................. 12–021
Howes Percival LLP v Page [2013] EWHC 4104 (Ch)............................... 9–039
Hunt v Optima (Cambridge) Ltd [2014] EWCA Civ 714; [2014] B.L.R. 613; 155 Con.
  L.R. 29; [2014] P.N.L.R. 29................................. 5–039, 6–027, 7–027, 16–017
Hurley Palmer Flatt Ltd v Barclays Bank Plc [2014] EWHC 3042 (TCC); [2014] B.L.R.
  713; 156 Con. L.R. 213.. 13–022
Inframatrix Investments Ltd v Dean Construction Ltd [2012] EWCA Civ 64; [2012] 2
  All E.R. (Comm) 337; 140 Con. L.R. 59; (2012) 28 Const. L.J. 438; [2012] C.I.L.L.
  3145.. 11–006, 16–014
Jackson v Dear; sub nom. Dear v Jackson [2012] EWHC 2060 (Ch)............... 3–002
JBW Group Ltd v Ministry of Justice [2012] EWCA Civ 8; 141 Con. L.R. 62; [2012] 2
  C.M.L.R. 10; [2012] Eu. L.R. 379..................................... 15–011, 15–045
Jet2.com Ltd v Blackpool Airport Ltd [2012] 142 Con. L.R. 1 (CA)............ 12–003
Jet2.com Ltd v SC Compania Nationala de Transporturi Aeriene Romane Tarom SA;
  sub nom. SC Compania Nationala de Transporturi Aeriene Romane Tarom SA v
  Jet2.com Ltd [2014] EWCA Civ 87... 6–066
Jim Ennis Construction Ltd v Premier Asphalt Ltd [2009] EWHC 1906 (TCC); 125
  Con. L.R. 141; [2009] 3 E.G.L.R. 7; [2009] 41 E.G. 116; [2009] C.I.L.L. 2745... 16–024
JJ Metcalfe v Dennison Unreported December 6, 2013 QBD (TCC)................ 16–014
John Grimes Partnership Ltd v Gubbins [2013] EWCA Civ 37; [2013] B.L.R. 126; 146
  Con. L.R. 26; [2013] P.N.L.R. 17.............................. 9–005, 9–006, 13–060

Table of Cases

Joyce v Rigolli [2004] EWCA Civ 79; (2004) 148 S.J.L.B. 234; [2004] 1 P. & C.R. DG22.. 2–034
Kingspan Environmental Ltd v Borealis A/S [2012] EWHC 1147 (Comm).......... 3–075
Kleinwort Benson Ltd v Lincoln City Council; Kleinwort Benson Ltd v Birmingham City Council; Kleinwort Benson Ltd v Southwark LBC; Kleinwort Benson Ltd v Kensington and Chelsea RLBC [1999] 2 A.C. 349; [1998] 3 W.L.R. 1095; [1998] 4 All E.R. 513; [1998] Lloyd's Rep. Bank. 387; [1999] C.L.C. 332; (1999) 1 L.G.L.R. 148; (1999) 11 Admin. L.R. 130; [1998] R.V.R. 315; (1998) 148 N.L.J. 1674; (1998) 142 S.J.L.B. 279; [1998] N.P.C. 145 HL.................................... 12–002
Laing O'Rourke Construction Ltd (formerly Laing O'Rourke Northern Ltd) v Healthcare Support (Newcastle) Ltd [2014] EWHC 2595 (TCC); [2014] B.L.R. 722; 155 Con. L.R. 148; [2014] C.I.L.L. 3549....................................... 5–014
Leander Construction Ltd v Mulalley & Co Ltd [2011] EWHC 3449 (TCC); [2012] B.L.R. 152; [2012] C.I.L.L. 3151................................ 3–043, 8–004, 9–052
Liberty Mercian Ltd v Cuddy Civil Engineering Ltd [2013] EWHC 4110 (TCC); [2014] C.I.L.L. 3469................................. 2–031, 3–024, 12–003, 12–012, 12–017A
Lictor Anstalt v Mir Steel UK Ltd; sub nom. Mir Steel UK Ltd v Morris [2012] EWCA Civ 1397; [2013] 2 All E.R. (Comm) 54; [2013] C.P. Rep. 7; [2013] 2 B.C.L.C. 76... 3–065
Lidl UK GmbH v RG Carter Colchester Ltd (2012) 146 Con. L.R. 133........... 10–006
Lincolnshire CC v Mouchel Business Services Ltd [2014] EWHC 352 (TCC); [2014] B.L.R. 347; [2014] T.C.L.R. 3; [2014] C.I.L.L. 3484.......................... 19–006
Lion Apparel Systems Ltd v Firebuy Ltd [2007] EWHC 2179 (Ch); [2008] Eu. L.R. 191.. 15–028
Lloyd v Browning [2013] EWCA Civ 1637; [2014] 1 P. & C.R. 11.................. 6–025
Lloyds TSB Foundation for Scotland v Lloyds Banking Group Plc [2013] UKSC 3; [2013] 1 W.L.R. 366; [2013] 2 All E.R. 103; 2013 G.W.D. 4–111............... 3–002
Lordsvale Finance Plc v Bank of Zambia [1996] Q.B. 752; [1996] 3 W.L.R. 688; [1996] 3 All E.R. 156; [1996] C.L.C. 1849 QBD...................................... 10–005
Lowry Brothers Ltd v Northern Ireland Water Ltd [2013] NIQB 23.............. 15–036
LW Infrastructure v Lim Chin San [2012] B.L.R. 13 (HC of Singapore).......... 10–020
Makdessi v Cavendish Square Holdings BV; sub nom. Cavendish Square Holdings BV v Makdessi; El Makdessi v Cavendish Square Holdings BV [2013] EWCA Civ 1539; [2014] 2 All E.R. (Comm) 125; [2013] 2 C.L.C. 968; [2014] B.L.R. 246.... 10–005
Manolete Partners Plc v Hastings BC [2014] EWCA Civ 562; [2014] 1 W.L.R. 4030; [2014] B.L.R. 389.. 16–031
Marley v Rawlings [2014] UKSC 2; [2014] 2 W.L.R. 213; [2014] 1 All E.R. 807; [2014] 2 F.L.R. 555; [2014] W.T.L.R. 299; 16 I.T.E.L.R. 642; [2014] Fam. Law 466; (2014) 158(4) S.J.L.B. 49... 3–028
McGeown v Direct Travel Insurance; sub nom. Direct Travel Insurance v McGeown [2003] EWCA Civ 1606; [2004] 1 All E.R. (Comm) 609; [2004] Lloyd's Rep. I.R. 599; (2003) 147 S.J.L.B. 1365.. 3–032
Mears Ltd v Leeds CC [2011] EWHC 1031 (TCC); [2011] Eu. L.R. 764........... 15–016
Mears Ltd v Shoreline Housing Partnership Ltd [2013] EWCA Civ 639; [2013] C.P. Rep. 39; [2013] B.L.R. 393; 148 Con. L.R. 221; [2013] C.I.L.L. 3388.......... 12–002
Mears Ltd v Shoreline Housing Partnership Ltd; sub nom. Shoreline Housing Partnership Ltd v Mears Ltd [2013] EWCA Civ 639; [2013] C.P. Rep. 39; [2013] B.L.R. 393; 148 Con. L.R. 221; [2013] C.I.L.L. 3388..................................... 2–032
Meritz Fire & Marine Insurance Co Ltd v Jan de Nul NV [2010] EWHC 3362 (Comm); [2011] 1 All E.R. (Comm) 1049; [2011] 1 C.L.C. 48; [2011] B.L.R. 320; [2011] T.C.L.R. 2; 134 Con. L.R. 252... 11–036
Mermec UK Ltd v Network Rail Infrastructure Ltd [2011] EWHC 1847 (TCC)..... 15–031, 15–032
Mid Essex Hospital Services NHS Trust v Compass Group UK and Ireland Ltd (t/a Medirest). See Compass Group UK and Ireland Ltd (t/a Medirest) v Mid Essex Hospital Services NHS Trust
Mileform Ltd v Interserve Security Ltd [2013] EWHC 3386 (QB)............. 2–021, 2–032
Miliangos v George Frank (Textiles) Ltd [1976] A.C. 443; [1975] 3 W.L.R. 758; [1975] 3 All E.R. 801; [1976] 1 Lloyd's Rep. 201; [1975] 2 C.M.L.R. 585; (1975) 119 S.J. 774 HL.. 19–076

TABLE OF CASES

Mir Steel UK Ltd v Morris. *See* Lictor Ansalt v Mir Steel UK Ltd
Mitchell v News Group Newspapers Ltd [2013] EWCA Civ 1537; [2014] 1 W.L.R. 795;
 [2014] 2 All E.R. 430; [2014] B.L.R. 89; [2013] 6 Costs L.R. 1008; [2014] E.M.L.R.
 13; [2014] C.I.L.L. 3452; (2013) 163(7587) N.L.J. 20. . . . . 19–055, 19–077
Montpellier Estates Ltd v Leeds CC [2013] EWHC 166 (QB). 15–020, 15–031, 15–032,
 15–045
Mortgage Express v Abensons Solicitors [2012] EWHC 1000 (Ch); [2012] 2 E.G.L.R.
 83; [2012] 27 E.G. 90; [2012] 18 E.G. 103 (C.S.). 16–019
MT Hojgaard A/S v E.ON Climate and Renewables UK Robin Rigg East Ltd [2014]
 EWHC 1088 (TCC); [2014] B.L.R. 450. 1–030
Mueller Europe Ltd v Central Roofing (South Wales) Ltd [2013] EWHC 237 (TCC);
 147 Con. L.R. 32. 9–060
Murray v Leisureplay Plc [2005] EWCA Civ 963; [2005] I.R.L.R. 946. 10–005
National Westminster Bank Plc v Kapoor [2011] EWCA Civ 1083; [2012] 1 All E.R.
 1201; [2011] B.P.I.R. 1680; [2011] N.P.C. 97; [2012] Bus. L.R. D25. 13–005
Nationwide Gritting Services Ltd v Scottish Ministers [2013] CSOH 119; 2013 G.W.D.
 26–523. 15–031
NATS (Services) Ltd v Gatwick Airport Ltd [2014] EWHC 3133 (TCC); [2014] B.L.R.
 697; 156 Con. L.R. 177. 15–036
Newbury v Sun Microsystems Ltd [2013] EWHC 2180 (QB); [2013] C.I.L.L. 3425. . . 2–020
Newcastle Upon Tyne Hospital NHS Foundation Trust v Newcastle Primary Care
 Trust [2012] EWHC 2093 (QB) 15–036
Nokia Corp v AU Optronics Corp [2012] EWHC 731 (Ch); [2012] U.K.C.L.R. 245. . 16–022
North v Brown [2012] EWCA Civ 223, CA. 13–004
Northern Ireland Housing Executive v Healthy Buildings Ltd [2014] NICA 27; 153
 Con. L.R. 87. 22–007
Northumbrian Water Ltd v Sir Robert McAlpine Ltd [2014] EWCA Civ 685; [2014]
 B.L.R. 605; 154 Con. L.R. 26; [2014] Env. L.R. 28. 11–038
Novoship (UK) Ltd v Mikhaylyuk; Novoship (UK) Ltd v Nikitin [2014] EWCA Civ
 908; [2014] W.T.L.R. 1521; (2014) 158(28) S.J.L.B. 37. 19–076
NP Aerospace v Ministry of Defence Unreported. 15–030
Nulty v Milton Keynes BC [2013] EWCA 15; [2013] B.L.R. 134. 9–058
Oakrock Ltd v Travelodge Hotels Ltd [2014] EWHC 1868 (TCC); [2014] B.L.R.
 593. 4–053
Obrascon Huarte Lain SA v Attorney General of Gibraltar [2014] EWHC 1028 (TCC);
 [2014] B.L.R. 484; [2014] C.I.L.L. 3536. . . . . 8–030, 22–013, 22–014
Oksana Mul v Hutton Construction Ltd [2014] EWHC 1797 (TCC); [2014] B.L.R. 529;
 154 Con. L.R. 159; [2014] C.I.L.L. 3529. 20–148
OMV Supply and Trading AG v Kazmunaygaz Trading AG (formerly Vector Energy
 AG) [2014] EWHC 1372 (Comm). 12–012
Page v Hewetts Solicitors (A Firm) [2012] EWCA Civ 805; [2012] C.P. Rep. 40; [2012]
 W.T.L.R. 1427. 16–013
Patel v Mirza [2014] EWCA Civ 1047; [2014] Lloyd's Rep. F.C. 561; [2014] W.T.L.R.
 1567. 6–051
PC Harrington Contractors Ltd v Systech International Ltd [2012] EWCA Civ 1371;
 [2013] 2 All E.R. 69; [2013] 1 All E.R. (Comm) 1074; [2013] Bus. L.R. 970; [2013]
 B.L.R. 1; 145 Con. L.R. 1; [2013] 1 E.G.L.R. 9; [2013] 3 E.G. 88; [2012] C.I.L.L.
 3260. 4–003
PCE Investors Ltd v Cancer Research UK [2012] 2 P. & C.R. 5 (Ch). 12–002
Pearson Driving Assessments Ltd v Minister for the Cabinet [2013] EWHC 2082
 (TCC). 15–030
Pegasus Management Holdings SCA v Ernst & Young [2012] EWHC 738 (Ch); [2012]
 2 B.C.L.C. 734; [2012] P.N.L.R. 24; [2012] S.T.I. 1387. . . . . 9–017, 13–021
Persimmon Homes (South Cost) Ltd v Hall Aggregates (South Coast) Ltd [2012]
 EWHC 2429 (TCC); [2012] C.I.L.L. 3265. 19–076
Persimmon Homes Ltd v Woodford Land Ltd [2012] B.L.R. 73 (Ch). 12–015A
PGF II SA v OMFS Co 1 Ltd [2013] EWCA Civ 1288; [2014] 1 W.L.R. 1386; [2014] 1
 All E.R. 970; [2014] C.P. Rep. 6; [2014] B.L.R. 1; 152 Con. L.R. 72; [2013] 6 Costs
 L.R. 973; [2013] 3 E.G.L.R. 16; [2013] 44 E.G. 98 (C.S.); (2013) 157(42) S.J.L.B.
 37. . . . . 19–008, 19–009

TABLE OF CASES

Phethean-Hubble v Coles [2012] EWCA Civ 349; [2012] R.T.R. 31.................. 9–062
Photo-me International Plc v Network Rail Infrastructure Ltd [2011] EWHC 3168
 (QB).. 15–011, 15–045
Pickard Finlason Partnership Ltd v Lock [2014] EWHC 25 (TCC)................. 14–039
Price v Carter [2010] 128 Con. L.R. 124.. 20–019
Prime Sight Ltd v Lavarello [2013] UKPC 22; [2014] A.C. 436; [2014] 2 W.L.R. 84;
 [2013] 4 All E.R. 659; (2014) 158(2) S.J.L.B. 37............................. 3–061
Qimonda Malaysia SDN BHD v Sediabena SDN BHD [2012] B.L.R. 65, CA
 (Mal).. 12–020, 13–006, 13–007
Quashie v Stringfellow Restaurants Ltd [2012] EWCA Civ 1735; [2013] I.R.L.R. 99;
 (2013) 157(1) S.J.L.B. 31... 13–054
R. (on the Application of All About Rights Law Practice) v Legal Services Commis-
 sion [2011] EWHC 964 (Admin)............................... 15–012, 15–022
R. (on the application of All About Rights Law Practice) v Lord Chancellor [2013]
 EWHC 3461 (Admin)... 15–019
R. (on the Application of Berky) v Newport CC [2012] W.L.R.(D) 128............ 15–042
R. (on the application of Castelli) v Merton LBC [2013] EWHC 602 (Admin)...... 16–072
R. (on the Application of Greenwich Community Law Centre) v Greenwich LBC
 [2011] EWHC 3463 (Admin); (2012) 156(1) S.J.L.B. 31...................... 15–019
R. (on the Application of Harrow Solicitors and Advocates) v Legal Services Com-
 mission [2011] EWHC 1087 (Admin); (2011) 108(20) L.S.G. 20; (2011) 155(18)
 S.J.L.B. 31; [2011] P.T.S.R. D49................................... 15–012, 15–022
R. (on the Application of Hoole and Co) v Legal Services Commission [2011] EWHC
 886 (Admin); [2011] Info. T.L.R. 1............................... 15–012, 15–022
R. (on the Application of Hossacks) v Legal Services Commission [2012] EWCA Civ
 1203; [2013] 1 Costs L.O. 94................................... 15–012, 15–022
R. (on the Application of Midlands Co-operative Society Ltd) v Birmingham CC and
 Tesco Stores Ltd [2012] EWHC 620 (Admin); [2012] Eu. L.R. 640; [2012]
 B.L.G.R. 393... 15–005
R. (on the application of Nash) v Barnet LBC [2013] EWCA Civ 1004; [2013] P.T.S.R.
 1457.. 15–042
R. (on the application of Pelling) v Newham LBC [2011] EWHC 3265 (Admin)..... 16–073
R. (on the application of Prudential Plc) v Special Commissioner of Income Tax [2013]
 UKSC 1; [2013] 2 W.L.R. 325; [2013] 2 All E.R. 247; [2013] S.T.C. 376; [2013] 2
 Costs L.R. 275; [2013] 1 F.C.R. 545; [2013] B.T.C. 45; [2013] C.I.L.L. 3309; [2013]
 S.T.I. 264; [2013] 5 E.G. 96 (C.S.); (2013) 163 N.L.J. 109..................... 19–050
R. (on the Application of UNISON) v NHS Wiltshire Primary Care Trust [2012]
 EWHC 624 (Admin); [2012] A.C.D. 84...................................... 15–042
R&C Electrical Engineers Ltd v Shaylor Construction Ltd [2012] EWHC 1254 (TCC);
 [2012] B.L.R. 373; 142 Con. L.R. 129; [2012] C.I.L.L. 3184................... 5–020
Rainy Sky SA v Kookmin Bank [2011] UKSC 50; [2011] 1 W.L.R. 2900; [2012] 1 All
 E.R. 1137; [2012] 1 All E.R. (Comm) 1; [2012] Bus. L.R. 313; [2012] 1 Lloyd's
 Rep. 34; [2011] 2 C.L.C. 923; [2012] B.L.R. 132; 138 Con. L.R. 1; [2011] C.I.L.L.
 3105.. 3–002
Rashid v Sharif [2014] EWCA Civ 377... 16–065
Regent ohG Aisestadt & Barig v Francesco of Jermyn Street Ltd [1981] 3 All E.R. 327;
 [1981] Com. L.R. 78 QBD.. 4–007
Relicpride Building Co Ltd v Cordara [2013] EWCA Civ 158; 147 Con. L.R. 92.... 4–005
Rendlesham Estates Plc v Barr Ltd [2014] EWHC 3968 (TCC).............. 16–002, 16–007
Resource (NI) Ltd v Ulster University [2013] NIQB 64........................... 15–018
Roche Diagnostics Ltd v Mid Yorkshire Hospitals NHS Trust [2013] EWHC 933
 (TCC); [2013] C.I.L.L. 3360; (2013) 163(7558) N.L.J. 17; [2013] P.T.S.R. D35... 15–030
Roe Brickwork Ltd v Wates Construction Ltd [2013] EWHC 3417 (TCC).......... 9–034
Rutledge Recruitment and Training Ltd v Department for Employment and Learning,
 Department of Finance and Personnel [2011] NIQB 61....................... 15–020
RWE Npower Renewables Ltd v JN Bentley Ltd [2014] EWCA Civ 150; [2014]
 C.I.L.L. 3488... 8–003
SABIC UK Petrochemicals Ltd (formerly Huntsman Petrochemicals (UK) Ltd) v Punj
 Lloyd Ltd; Punj Lloyd Ltd v SABIC UK Petrochemicals Ltd (formerly Huntsman

TABLE OF CASES

Petrochemicals (UK) Ltd) [2013] EWHC 2916 (QB); [2014] B.L.R. 43; [2013] Bus. L.R. D81. 20–056
Sainsbury's Supermarkets Ltd v Condek Holdings Ltd (formerly Condek Ltd) [2014] EWHC 2016 (TCC); [2014] B.L.R. 574; [2014] T.C.L.R. 5; [2014] C.I.L.L. 3553. 19–006
SC Compania Nationala de Transporturi Aeriene Romane Tarom SA v Jet2.com Ltd. See Jet2.com Ltd v SC Compania Nationala de Transporturi Aeriene Romane Tarom SA
Scottish Widows Fund and Life Assurance Society v BGC International (formerly Cantor Fitzgerald International) [2012] EWCA Civ 607; [2012] 142 Con. L.R. 27 (CA). 3–011, 3–022, 12–012
Seele Middle East FZE v Drake and Scull International SA Co [2014] EWHC 435 (TCC). 12–020
Seele Middle East FZE v Raiffeisenlandesbank Oberosterreich Aktiengesellschaft Bank [2014] EWHC 343 (TCC). 12–020
Sevcon Ltd v Lucas CAV Ltd [1986] 1 W.L.R. 462; [1986] 2 All E.R. 104; [1986] F.S.R. 338; [1986] R.P.C. 609; (1986) 83 L.S.G. 1641; (1986) 130 S.J. 340 HL. 16–014
Shetland Line (1984) Ltd v Scottish Ministers [2012] CSOH 99; 2012 G.W.D. 24–504. .... 15–028, 15–036
Simon Carves Ltd v Ensus UK Ltd [2011] EWHC 657 (TCC); [2011] B.L.R. 340; 135 Con. L.R. 96. 12–020
Simpson v Norfolk and Norwich University Hospital NHS Trust [2011] EWCA Civ 1149; [2012] Q.B. 640; [2012] 2 W.L.R. 873; [2012] 1 All E.R. 1423; [2012] 1 Costs L.O. 9; [2012] P.I.Q.R. P2; (2012) 124 B.M.L.R. 1; (2011) 108(41) L.S.G. 24; (2011) 161 N.L.J. 1451. 13–020
Sita UK Ltd v Greater Manchester Waste Disposal Authority [2011] EWCA Civ 156; [2012] P.T.S.R. 645; [2011] T.C.L.R. 3; 134 Con. L.R. 1; [2011] 2 C.M.L.R. 32; [2011] Eu. L.R. 719; [2011] B.L.G.R. 419; (2011) 108(10) L.S.G. 22. 15–031
Skywell (UK) Ltd v Revenue and Customs Commissioners [2012] UKFTT 61 (TC). 13–020
Smales v Lea [2011] EWCA Civ 1325; 140 Con. L.R. 70; [2012] P.N.L.R. 8. .... 4–003, 9–047
Spiers Earthworks v Landtec Projects [2012] B.L.R. 223 (WASC). 10–005
Squibb Group Ltd v Vertase FLI Ltd [2012] EWHC 1958 (TCC); [2012] B.L.R. 408; [2013] B.L.R. 11; [2012] C.I.L.L. 3220. 13–058
Steel Co of Canada Ltd v Willand Management Ltd [1966] S.C.R. 746 Sup Ct (Can). 1–030, 3–054
Straw Realisations (No.1) Ltd v Shaftesbury House Developments Ltd [2010] All E.R. (d) 196. 20–374
Suisse Atlantique v NV Rotterdamsche [1967] A.C. 361 HL. 10–020
Swallowfalls Ltd v Monaco Yachting and Technologies SAM [2014] EWCA Civ 186; [2014] 2 All E.R. (Comm) 185; [2014] 2 Lloyd's Rep. 50. .... 3–042, 3–046
Sycamore Bidco Ltd v Breslin [2013] EWHC 174 (Ch). 4–021
Tachie v Welwyn Hatfield BC [2013] EWHC 3972 (QB); [2014] P.T.S.R. 662; [2014] B.L.G.R. 100. 15–011
Tallington Lakes Ltd (No.01595671) v Larking Gowen (A Firm) [2014] EWCA Civ 959. 4–020
Taylor v Bowers (1876) 1 Q.B.D. 291 CA. 6–051
Ted Baker Plc v AXA Insurance UK Plc [2012] EWHC 1406 (Comm); [2013] 1 All E.R. (Comm) 129; [2013] Lloyd's Rep. I.R. 174; (2012) 109(28) L.S.G. 21. 3–007
Test Claimants in the FII Group Litigation v Revenue and Customs Comissioners [2012] UKSC 19; [2012] 2 A.C. 337; [2012] 2 W.L.R. 1149; [2012] 3 All E.R. 909; [2012] Bus. L.R. 1033; [2012] S.T.C. 1362; [2012] B.T.C. 312; [2012] S.T.I. 1707. 16–019
Thameside Construction Co Ltd v Stevens [2013] EWHC 2071 (TCC); [2013] B.L.R. 543; 149 Con. L.R. 195; [2013] C.I.L.L. 3392; [2013] Bus. L.R. D59. 13–058
Thinc Group v Armstrong [2012] EWCA Civ 1227. 3–018
Thomas Construction Ltd v Hyland (2011) C.I.L.L. 1748 QBD TCC. 19–006
Tinseltime Ltd v Roberts [2011] EWHC 1199; [2011] B.L.R. 515. 11–040

TABLE OF CASES

Tinsley v Milligan [1994] 1 A.C. 340; [1993] 3 W.L.R. 126; [1993] 3 All E.R. 65; [1993] 2 F.L.R. 963; (1994) 68 P. & C.R. 412; [1993] E.G. 118 (C.S.); [1993] N.P.C. 97 HL. 6–051
Traffic Signs and Equipment Ltd v Department for Regional Development [2010] NIQB 138. .... 15–032, 15–042
Trebor Bassett Holdings Ltd v ADT Fire & Security Plc [2012] EWCA Civ 1158; [2012] B.L.R. 441; 144 Con. L.R. 1. 3–052
Tribe v Tribe [1996] Ch. 107; [1995] 3 W.L.R. 913; [1995] 4 All E.R. 236; [1995] C.L.C. 1474; [1995] 2 F.L.R. 966; [1996] 1 F.C.R. 338; (1996) 71 P. & C.R. 503; [1996] Fam. Law 29; (1995) 92(28) L.S.G. 30; (1995) 145 N.L.J. 1445; (1995) 139 S.J.L.B. 203; [1995] N.P.C. 151; (1995) 70 P. & C.R. D38 CA (Civ Div). 6–051
Trustees of Ampleforth Abbey Trust v Turner & Townsend Management Ltd [2012] EWHC 2137 (TCC); [2012] T.C.L.R. 8; 144 Con. L.R. 115; [2012] C.I.L.L. 3252. 2–008
Turning Point Ltd v Norfolk CC [2012] EWHC 2121 (TCC); [2012] Eu. L.R. 800... 15–031, 15–032, 15–045
Turriff Ltd v Richards & Wallington (Contracts) Ltd [1981] Com. L.R. 39; 18 B.L.R. 13 QBD. 6–069
Unaoil Ltd v Leighton Offshore Pte Ltd [2014] EWHC 2965 (Comm); 156 Con. L.R. 24. 10–005
United International Pictures v Cine Bes Filmcilik ve Yapimcilik AS; sub nom. Cine Bes Filmcheck ve Yapimcilik AS v United International Pictures [2003] EWCA Civ 1669; [2004] 1 C.L.C. 401; (2003) 147 S.J.L.B. 1396. 10–005
University of Brighton v Dovehouse Interiors Ltd. *See* Brighton University v Dovehouse Interiors Ltd
Urban I (Blonk Street) Ltd v Ayres [2013] EWCA Civ 816; [2014] 1 W.L.R. 756; [2013] B.L.R. 505; [2014] 1 P. & C.R. 1; [2013] 3 E.G.L.R. 91; [2013] 29 E.G. 105 (C.S.). 6–077
Valilas v Januzaj [2014] EWCA Civ 436; 154 Con. L.R. 38. 6–060
Walker Construction (UK) Ltd v Quayside Homes Ltd [2014] EWCA Civ 93; [2014] 1 C.L.C. 121; [2014] B.L.R. 215; 153 Con. L.R. 26; [2014] 3 Costs L.O. 362. ..... 3–040, 16–024
Walter Lilly & Co Ltd v Mackay [2012] EWHC 1773 (TCC); [2012] B.L.R. 503; 143 Con. L.R. 79; (2012) 28 Const. L.J. 622; [2012] C.I.L.L. 3229. 503. .... 8–026, 8–027, 8–036, 8–062, 8–063, 8–066, 9–033, 9–041, 9–059, 19–021, 19–050, 20–100, 20–120, 20–273
Walter Lilly & Co Ltd v Mackay [2012] EWHC 1773 (TCC); [2012] B.L.R. 503; 143 Con. L.R. 79; (2012) 28 Const. L.J. 622; [2012] C.I.L.L. 3229. 8–037
West Country Renovations Ltd v McDowell [2012] EWHC 307 (TCC); [2013] 1 W.L.R. 416; [2012] 3 All E.R. 106; [2012] B.L.R. 255; 141 Con. L.R. 112; [2012] C.I.L.L. 3158. 19–014
West v Ian Finlay and Associates [2014] EWCA Civ 316; [2014] B.L.R. 324; 153 Con. L.R. 1; [2014] C.I.L.L. 3507. .... 3–078, 9–053, 9–062
Westcoast (Holdings) Ltd (formerly Kelido Ltd) v Wharf Land Subsidiary (No.1) Ltd [2012] EWCA Civ 1003. 3–034
Westshield Ltd v Whitehouse [2013] EWHC 3576 (TCC); [2014] Bus. L.R. 268; [2014] B.P.I.R. 317; [2014] C.I.L.L. 3457. 16–040
Westvilla Properties Ltd v Dow Properties Ltd [2010] EWHC 30 (Ch); [2010] 2 P. & C.R. 19; [2010] 2 P. & C.R. DG4. 3–004
Woodlands Oak Ltd v Conwell [2011] EWCA Civ 254; [2011] B.L.R. 365. 9–046
WS Tankship II BV v Kwanju Bank Ltd [2011] EWHC 3103; [2012] C.I.L.L. 3155. 11–036
Wuhan Guoyo Logistics Group Co Ltd v Emporiki Bank of Greece SA [2012] EWCA Civ 1629; [2013] 1 All E.R. (Comm) 1191; [2012] 2 C.L.C. 986; [2013] B.L.R. 74; [2013] C.I.L.L. 3300. .... 11–035, 11–036
Wuhan Ocean Economic & Technical Cooperation Co Ltd v Schiffahrts-Gesellschaft Hansa Murcia mbH & Co KG [2012] EWHC 3104 (Comm); [2013] 1 All E.R. (Comm) 1277; [2013] 1 Lloyd's Rep. 273. 3–043
WW Gear v McGee Group [2012] 1 B.L.R. 355. 20–273
XY v Facebook Ireland Ltd [2012] NIQB 96. 12–017

Yam Seng Pte Ltd v International Trade Corp Ltd [2013] EWHC 111 (QB); [2013] 1 All E.R. (Comm) 1321; [2013] 1 Lloyd's Rep. 526; [2013] B.L.R. 147; 146 Con. L.R. 39.. 9–014
Yeates v Line [2012] EWHC 3085 (Ch); [2013] Ch. 363; [2013] 2 W.L.R. 844; [2013] 2 All E.R. 84; [2013] 1 P. & C.R. 22; [2012] 47 E.G. 126 (C.S.); [2013] 1 P. & C.R. DG7.. 2–034
Zenstrom v Fagot [2013] EWHC 288 (TCC); 147 Con. L.R. 162.................. 16–004

TABLE OF EUROPEAN CASES

Alfastar Benelux SA v Council of the European Union (T-57/09) [2011] OJ C355/16. 15–029
Azienda Sanitaria Locale di Lecce v Ordine degli Ingegneri della Provincia di Lecce (C–159/11) [2013] P.T.S.R. 1043 ; [2013] 2 C.M.L.R. 17. 15–011
Bundesdruckerei GmbH v Stadt Dortmund (C-549/13) [2014] I.R.L.R. 980. 15–016
Centro Hospitalar de Setubal EPE v Eurest (Portugal) - Sociedade Europeia de Restaurantes Lda (C-574/12) [2014] P.T.S.R. 1174; (2014) 158(26) S.J.L.B. 37. 15–011
Commission of the European Communities v France (C-237/99) [2001] E.C.R. I–939. 15–009
Commission of the European Communities v Netherlands (C-368/10) [2013] All E.R. (EC) 804; [2012] 3 C.M.L.R. 11. 15–016
Consorzio Stabile Libor Lavori Pubblici v Comune di Milano (C-358/12) ECLI:EU:C:2014:2063 July 10, 2014. 15–013
Datenlotsen Informationssysteme GmbH v Technische Universität Hamburg-Harburg (C-15/13). *See* Technische Universität Hamburg-Harburg v Datenlotsen Informationssysteme GmbH
Econord SpA v Commune di Cagno (C-182/11) [2013] P.T.S.R. 356; [2013] 2 C.M.L.R. 7. 15–011
Evropaïki Dynamiki v Commission of the European Communities (T-39/08) [2012] OJ C25/47. . . . . 15–016, 15–029
Evropaïki Dynamiki v Court of Justice of the European Union (T-447/10) [2012] OJ C373/5. 15–016
Forposta SA v Poczta Polska SA (C-465/111) [2013] OJ C38/8. 15–013
Idrodinamica Spurgo Velox srl v Acquedotto Pugliese SpA (C-161/13) [2014] P.T.S.R. 935. 15–022
Impresa Pizzarotti & C. SpA v Comune di Bari (C-213/13) ECLI:EU:C:2014:2067 July 10, 2014. 15–005
Insinööritoimisto InsTiimi Oy (C-615/10) [2012] OJ C217/3. 15–002
IVD GmbH & Co KG v Arztekammer Westfalen-Lippe (C-526/11) [2014] 1 C.M.L.R. 31; [2014] C.E.C. 453. 15–009
IVD GmbH & Co. KG v Ärztekammer Westfalen–Lippe (C-526/11) [2012] OJ C25/28. 15–009
Ministeriet for Forskning, Innovation og Videregaende Uddannelser v Manova A/S (C-336/12) [2014] P.T.S.R. 254. 15–019
Ministero dell'Interno v Fastweb SpA (C-19/13) ECLI:EU:C:2014:2194 September 11, 2014. 15–038
Nordecon AS v Rahandusministeerium (C-561/12) [2014] P.T.S.R. 343. 15–028
Piepenbrock Dienstleistungen GmbH & Co KG v Kreis Düren (C-386/11) [2013] OJ C225/7. 15–011
SAG ELV Slovensko as v Úrad pre verejné obsarávanie (C-599/10) [2013] P.T.S.R. 1; [2012] 2 C.M.L.R. 36. . . . . 15–018, 15–022
Serco Belgium SA v European Commission (T-644/13). 15–028
SIAC Construction Ltd v Mayo CC (C-19/00) [2001] E.C.R. I-7725; [2002] T.C.L.R. 16; [2001] 3 C.M.L.R. 59; [2002] All E.R. (EC) 272. 15–016
Star Fruit Co SA v Commission of the European Communities (C-247/87) [1989] E.C.R. 291; [1990] 1 C.M.L.R. 733. 15–043

TABLE OF EUROPEAN CASES

Technische Universität Hamburg-Harburg v Datenlotsen Informationssysteme GmbH
    (C-15/13) [2014] P.T.S.R. 1293... 15–011
Uniplex (UK) Ltd v NHS Business Services Authority (C-406/08) [2010] P.T.S.R.
    1377; [2010] 2 C.M.L.R. 47... 15–031, 15–042

TABLE OF STATUTES

1838	Judgments Act (c.110)–			(1)...	15–046
	s.17...	19–076		(2)...	15–046
1940	Law Reform (Miscellaneous Provisions) (Scotland) Act (c.42)...	3–073		(4)...	15–046
				(6)...	15–046
				s.82...	15–046
1967	Misrepresentation Act (c.7)..	20–405		s.83(1)...	15–046
1970	Administration of Justice Act (c.31)–			(2)...	15–046
				(3)...	15–046
	s.44A...	19–076		(8)...	15–046
1978	Civil Liability (Contribution) Act (c.47)...	3–073		(11)...	15–046
			2012	Public Services (Social Value) Act (c.3)...	15–046
1981	Senior Courts Act (c.54)...	19–010			
	s.19...	12–107A		s.1(1)(a)...	15–046
	s.35A...	19–076		(b)...	15–046
1984	Building Act (c.55)–			(2)...	15–046
	s.106(1)...	16–031		(3)(a)...	15–046
1996	Housing Grants, Construction and Regeneration Act (c.53)–			(b)...	15–046
				(4)...	15–046
				(7)...	15–046
	Ch.II...	3–041		(11)...	15–046
2008	Planning Act (c.29)...	15–005		(13)...	15–046
2009	Banking Act (c.1)...	16–032		s.4(4)...	15–046
2011	Localism Act (c.20)...	15–046	2012	Legal Aid, Sentencing and Punishment of Offenders Act (c.10)	
	Pt 2...	15–046			
	s.48...	15–046			
	s.51...	15–046		s.45...	19–077
	s.52...	15–046	2013	Growth and Infrastructure Act (c.27)...	15–005
	s.81...	15–046			

TABLE OF STATUTORY INSTRUMENTS

2006	Public Contracts Regulations (SI 2006/5)......	15–002, 15–042	2010	Building (Approved Inspectors etc) Regulations (SI 2010/2215)............. 16–025
	reg.1(1)	15–002	2011	Defence and Security Public Contracts Regulations (SI 2011/1848)......... 15–042
	reg.4(3)..................	15–001		
	reg.47D(2)...............	15–042		
	reg.47L	15–038		
	reg.49(1).................	15–002	2011	Public Procurement (Miscellaneous Amendments) Regulations (SI 2011/2053)........... 15–002, 15–027
2006	Utilities Contracts Regulations (SI 2006/6).......	15–002, 15–042		
	reg.1(1)	15–002	2012	Localism Act 2011 (Commencement No. 5 and Transitional, Savings and Transitory Provisions) Order (SI 2012/1008)................. 15–046
	reg.4(3)..................	15–001		
	reg.48(1)	15–002		
2007	Public Contracts and Utilities Contracts (Amendment) Regulations (SI 2007/3542).................	15–002	2012	Localism Act 2011 (Commencement No. 6 and Transitional, Savings and Transitory Provisions) Order (SI 2012/1463)................. 15–046
2009	Public Contracts (Amendment) Regulations (SI 2009/2992).............	15–002		
2009	Utilities Contracts (Amendment) Regulations (SI 2009/3100).............	15–002	2012	Community Right to Challenge (Fire and Rescue Authorities and Rejection of Expressions of Interest) (England) Regulations (SI 2012/1647) Sch.1.................... 15–046
2010	Legislative Reform (Insolvency) (Miscellaneous Provisions) Order (SI 2010/18)...............	16–032		
2010	Building (Local Authority Charges) Regulations (SI 2010/404)..............	16–028	2012	Public Services (Social Value) Act 2012 (Commencement) Order (SI 2012/3173).................. 15–046
2010	Insolvency (Amendment) Rules (SI 2010/686).....	16–032		
2010	Insolvency (Amendment) (No.2) Rules (SI 2010/734)...................	16–032		

TABLE OF CIVIL PROCEDURE RULES

1998 Civil Procedure Rules (SI 1998/3132)	
Pt 1.	19–001
r.1.4.2(e).	1–038
Pt 3.	19–001
r.3.1.	19–077
r.3.9(1).	19–055
rr.3.12–3.18.	19–077
r.3.13.	19–077
r.3.14.	19–077
r.3.18.	19–077
Pt 29.4.	19–012
Pt 31.	19–001
r.31.5(3).	19–049
(5).	19–049
(6).	19–049
(7).	19–049
r.32.2(3).	19–043
Pt 35.	19–001
r.35.4(3).	19–045
Pt 35 PD para.11.4.	19–046
Pt 36.	19–001, 19–077, 19–088
r.36.14(1A).	19–088
Pt 43.	19–077
Pts 43–48.	19–001
r.44.2.	19–082, 19–090
(1).	19–080
(2)(a).	19–080
(4).	19–009, 19–080
(c).	19–088
(5)(a).	19–006
(6)(g).	19–076
r.44.3(1)(a).	19–080
(2)(a).	19–077
(5).	19–077
r.44.4.	19–077
r.44.9(4).	19–076
r.44.11.	19–082
r.46.8.	19–082
r.54.5(6).	15–042
2013 Civil Procedure (Amendment) Rules (SI 2013/262)	
r.14.	19–077
r.22(7).	19–077
2013 Damages–Based Agreements Regulations (SI 2013/609).	19–077

TABLE OF EUROPEAN LEGISLATION

Treaties
2010 Treaty on the Functioning of the European Union (TFEU)
art.258.................. 15–043
art.260.................. 15–046
(2).................... 15–043

Directives
1989 Dir.89/665/EEC on the coordination of the laws, regulations and administrative provisions relating to the application of review procedures to the award of public supply and public works contracts [1989] OJ L395/33.................... 15–038
2004 Dir.2004/17/EC on procurement procedures in the water, energy, transport and postal services sectors [2004] OJ L134/1... 15–002
art.10.................. 15–001
2004 Dir.2004/18/EC on procedures for the award of public works contracts, public supply contracts and public service contracts [2004] OJ L134/114........... 15–002, 15–016
art.2.................... 15–001
2004 Dir.2004/23/EC on setting standards of quality and safety for the donation, procurement, testing, processing, preservation, storage and distribution of human tissues and cells [2004] OJ L102/48.................... 15–002
2004 Dir.2004/25/EC on takeover bids [2004] OJ L142/12.................... 15–002
2007 Dir.2007/66/EC amending Council Directives 89/665/EEC and 92/13/EEC with regard to improving the effectiveness of review procedures concerning the award of public contracts [2007] OJ L335/31............ 15–038
2014 Dir.2014/24/EU on public procurement [2014] OJ L94/65......... 15–001, 15–002, 15–012, 15–014, 15–015, 15–018, 15–025
Recital (28).............. 15–012
Recital (36).............. 15–012
art.18................... 15–001
(2).................... 15–018
art.31................... 15–014
art.36(2)................. 15–018
art.49................... 15–014
art.53................... 15–025
art.67................... 15–015
art.69(3)................. 15–018
art.73................... 15–025
arts 74–77............... 15–012
art.82................... 15–015
art.84(3)................. 15–018
arts 91–94............... 15–012
Annex XVII............. 15–012
2014 Dir.2014/55/EU on electronic invoicing in public procurement [2014] OJ L133/1................ 15–025

TABLE OF REFERENCES TO THE JCT STANDARD FORM OF BUILDING CONTRACT

2011 Standard Form of Building Contract
- cl.1.3.................... 20–029
- cl.4.9.................... 20–029
- cl.4.10................... 20–029
- cl.8.5.3.1................ 20–374
- cl.8.7.3.................. 20–374
- cl.17.1................... 20–120
- cl.25..................... 20–100
- cl.25.3.1................. 20–100
- cl.25.3.3........... 8–036, 20–100
- cl.26..................... 20–273
- cl.26.1................... 20–273
- cl.26.2................... 20–273

Chapter One

THE NATURE OF A CONSTRUCTION CONTRACT

1.	Introduction	1–001
2.	The persons concerned in a construction contract	1–003
3.	Contract documents	1–014
☐ 4.	Contractual arrangements	1–022
5.	A typical construction operation—traditional procedure	1–025
☐ 6.	Design and build contracts	1–027
7.	Management and construction management contracts	1–033
8.	Contracts with financial and operating obligations	1–034
9.	Smaller works contracts	1–035
☐ 10.	Dispute resolution	1–036

4. CONTRACTUAL ARRANGEMENTS

[Add new sentence at end of note 26: page 7]   1–022

Since the First Supplement to the Ninth Edition, of July 2013, the Restructuring Group (established by the Users Forum of the sponsoring bodies, the ACE and CECEA) has produced a new and substantially revised version of the ICC Conditions of Contract, drafted so as to be suitable for use with a wider range of construction activities and for use either within the UK or abroad. The new draft was approved by its sponsors and was published in October 2014. It is intended that the new draft will be used as a template for updating other forms within the existing suite of ICC documents, all of which (with the exception of the "blue form" of sub-contract) have been inherited from the ICE as noted in the main work. It is likely that the next documents to be produced will be a sub-contract form and a target cost version of the main contract form.

[Substitute note 27: page 7]

There is also the NEC3 Engineering and Construction Subcontract (ECS); Short Contract (ECSC); Short Subcontract (ECSS); NEC3 Professional Services Contract (PSC); NEC3 Professional Services Short Contract (PSSC); NEC3 Term Service Contract (TSC); NEC3 Term Service Short Contract (TSSC); NEC3 Supply Contract (SC); NEC3 Supply Short Contract (SSC); NEC3 Framework Contract (FC); and NEC3 Adjudicator's Contract (AC). All of the NEC 3 contracts were issued in an amended form on April 1, 2013.

[Add new note 29A at end of final paragraph of 1–022 ("A brief commentary on NEC3 is included in Ch.22."): page 7]

[29A] For a detailed consideration of NEC3, see D. Thomas QC, *Keating on NEC*, 1st edn (London: Sweet & Maxwell, 2012).

1–023 [Substitute second sentence of note 31: page 7]

These now include: *Form of Contract—The Red Book—Lump Sum Contracts 5th edition—UK Version* (London: IChemE, 2013), *Form of Contract—The Green Book—Reimbursable Contracts 4th Edition—UK Version* (London: IChemE, 2013), *Form of Contract—The Burgundy Book 2nd Edition—Target Cost Contract—UK Version* (London: IChemE, 2013), *Form of Contract—The Yellow Book 4th Edition—Subcontracts—UK Version* (London: IChemE, 2013), *Form of Contract—The Brown Book 3rd Edition—Subcontract for Civil Engineering Works—UK Version* (London: IChemE, 2013), and *Form of Contract Minor Works, 2nd edition—The Orange Book* (London: IChemE, 2003).

[Insert after final sentence in note 31: page 7]

The RIBA Agreements all now have a 2012 Revision.

1–024 [Add new sentence at end of note 37: page 8]

The ACE Agreements are now in their Second Revision.

[Substitute fifth sentence of third paragraph of 1–024: page 9]

FIDIC also produce Conditions of Contract for Design, Build and Turnkey[39] (1995), Electrical and Mechanical Works[40] (1987) and Design, Build and Operate Projects (2008).[40A]

[40A] Known as the *Gold Book*.

[Add new sentence at beginning of note 41: page 9]

The New FIDIC Contracts were issued in 1999 and a memorandum was issued on April 1, 2013 setting out amendments to the Red, Yellow and Silver Books in relation to the enforcement of DAB decisions that are binding and not yet final.

6. DESIGN AND BUILD CONTRACTS

Standard Contract Forms.

[Substitute second sentence of paragraph 1–028] 1–028

The JCT publish the Design and Build Contract (2011) and the Standard Form of Building Contract, which incorporates a Contractor's Designed Portion option, whilst there is a provision in the Infrastructure Conditions of Contract[55] and ACE/CECA also publish a separate Infrastructure Conditions of Contract Design and Construct Version.

[Substitute note 61: page 13] 1–030

[61] *Steel Co of Canada Ltd v Willand Management Ltd* [1966] S.C.R. 746 (Canada SC); considered in *MT Hojgaard A/S v E.ON Climate and Renewables UK Robin Rigg East Ltd* [2014] B.L.R. 450; [2014] EWHC 1088 (TCC).

10. DISPUTE RESOLUTION

[Substitute note 77: page 17] 1–038

[77] See CPR r.1.4.2(e) and S. Blake, J. Browne, and S. Sime, *The Jackson ADR Handbook*, 1st edn (Oxford: Oxford University Press, 2013).

[Substitute note 78]

[78] See paras 7.5 and 7.6 of the *TCC Guide* (2nd edn, 3rd revision, London, April 2014).

CHAPTER TWO

FORMATION OF CONTRACT

☐ 1.	Elements of Contract		2–001
2.	Offer and acceptance		2–002
	(a)	*Invitation to tender*	2–002
☐	(b)	*Tender*	2–003
☐	(c)	*Letter of intent*	2–007
	(d)	*Estimates*	2–009
	(e)	*Standing offers*	2–010
☐	(f)	*Rejection and revocation of offer*	2–011
☐	(g)	*Unconditional acceptance*	2–016
■	(h)	*Acceptance by conduct*	2–027
■	(i)	*Acceptance by post, telex, fax or email*	2–028
	(j)	*Notice of terms*	2–029
	(k)	*Course of dealing*	2–030
3.	Formalities of contract		2–033
☐	(a)	*Contracts requiring writing*	2–034
	(b)	*Deeds*	2–036
4.	Capacity of parties		2–037
	(a)	*Minors (or infants)*	2–038
☐	(b)	*Aliens*	2–039
	(c)	*Bankrupts*	2–040
	(d)	*Persons of unsound mind*	2–041
	(e)	*Corporations*	2–042
	(f)	*Contracts by local authorities*	2–043
	(g)	*Government contracts*	2–045
	(h)	*Partnerships*	2–046
☐	(i)	*Unincorporated associations*	2–047

CHAPTER TWO—FORMATION OF CONTRACT

1. ELEMENTS OF CONTRACT

2–001 [Substitute first sentence of note 2: page 19]

See *Chitty on Contracts*, edited by H. Beale, 31st edn (London: Sweet & Maxwell, 2012), Vol.1, Ch.2.

[Amend reference in note 9: pages 19 and 20]

Chitty on Contracts, edited by H. Beale, 31st edn (London: Sweet & Maxwell, 2012), Vol.1, Ch.5.

2. OFFER AND ACCEPTANCE

(b) Tender

2–003 [Amend reference in note 18: page 21]

see *Chitty on Contracts*, edited by H. Beale, 31st edn (London: Sweet & Maxwell, 2012), Vol.1, para. 21–084.

2–004 [Amend references in note 25: page 22]

Lord Goff of Chieveley and G. Jones, *Goff and Jones: The Law of Unjust Enrichment*, 8th edn (London: Sweet & Maxwell, 2011), Ch.16; *Chitty on Contracts*, edited by H. Beale, 31st edn (London: Sweet & Maxwell, 2012), Vol.1, paras 29–020 and 29–076;

(c) Letter of intent

Documents so described are frequently sent.

2–007 [Add to the end of paragraph 2–007: page 23]

If the intended future contract is not made, a professional charged with procuring it, may be liable in negligence. In *Trustees of Ampleforth Abbey Trust v Turner & Townsend Management Ltd*,[36A] the defendant project managers' decision to proceed with the works under a series of letters of intent amounted to a failure to exercise reasonable care and skill. Had the intended future contract been executed, there was a real and substantial chance that the claimant employer would have been entitled to liquidated damages for delay to the works caused by the building contractor.

[36A] (2012) 144 Con. L.R. 115; [2012] EWHC 2137 (TCC).

(f) Rejection and revocation of offer

Rejection.

[Amend reference in note 50: page 26]   2–011

See also, *Chitty on Contracts*, edited by H. Beale, 31st edn (London: Sweet & Maxwell, 2012), Vol.1, para.2–092.

Counter-offer.

[Amend reference in note 51: page 26]   2–012

and see *Chitty on Contracts*, edited by H. Beale, 31st edn (London: Sweet & Maxwell, 2012), Vol.1, paras 2–032 and 2–092.

(g) Unconditional acceptance

"Battle of forms".

[Substitute note 69: page 28]   2–017

[69] *Tekdata v Amphenol* [2009] EWCA Civ 1209, [2010] Lloyd's Rep. 357, Dyson L.J., [25]; applied in *Trebor Bassett Holdings v ADT Fire and Security* [2011] EWHC 1936 (TCC), [2011] B.L.R. 661, [157].

Negotiations for a contract.

[Amend reference in note 74: page 29]   2–018

Chitty on Contracts, edited by H. Beale, 31st edn (London: Sweet & Maxwell, 2012), Vol.1, para.2–065.

"Subject to contract".

[Add to text at the end of note 89: page 32]   2–020

The words "such settlement to be recorded in a suitably worded agreement" may not be a condition precedent to a concluded contract but may indicate that the terms of the settlement agreement would be committed to writing as a record of what had already been agreed: *Newbury v Sun Microsystems* [2013] EWHC 2180 (QB), [21].

Certainty of terms.

[Add sentence to end of note 102: page 34]   2–021

See also *Mileform Ltd v Interserve Security Ltd* [2013] EWHC 3386 (QB)—an "exclusivity" term was too uncertain to form part of any agreement.

[Substitute last sentence of note 103: page 34]

For the effect of the words "to be agreed" on the certainty of terms and the ability of the court to supply certainty, see *Mamidoil-Jetoil Greek Petroleum v Okta Crude Oil Refinery* [2001] 2 Lloyd's Rep. 76, 89, CA; *Willis Management Ltd v Cable & Wireless Plc* [2005] 2 Lloyd's Rep. 597, 604, CA; and *MRI Trading AG v Erdenet Mining Corporation LLC* [2013] EWCA Civ 156, CA (where a shipping schedule and certain terms were to "be agreed during the negotiations of terms for 2010 terms").

[Add new fourth paragraph to 2–021: page 34]

A contractual obligation to use best endeavours may not be so uncertain that it is incapable of giving rise to a legally binding obligation and should usually be held to be an enforceable obligation and given practical content. However, it might be difficult to define the precise limits of the obligation in advance and in any given case to determine whether the term had been breached.[105A] Such contractual provisions cannot be enforced if the object of an obligation to use best endeavours is too vague or elusive to be itself a matter of legal obligation or if the parties have provided no criteria to assess whether best endeavours have been or can be used.[105B]

[105A] per Moore-Bick L.J. in a majority (Lewison L.J. dissenting) in *Jet2.Com Limited v Blackpool Airport Limited* [2012] EWCA Civ 417; (2012) 142 Con. L.R. 1, CA, [29] in relation to an obligation to use best endeavours to promote another person's business.

[105B] per Longmore L.J. in a majority (Lewison L.J. dissenting) in *Jet2.Com Limited v Blackpool Airport Limited* [2012] EWCA Civ 417, (2012) 142 Con. L.R. 1, CA, [69] adopting the words of Potter L.J. in *Phillips Petroleum v Enron Europe Ltd* [1997] C.L.C. 329, 343, CA. See also *Dany Lions Ltd v Bristol Cars Ltd* [2014] EWHC 817 (QB), where an obligation to use "reasonable endeavours" to enter into an agreement with a third party was unenforceable as there was no prior identification of terms to be included in that agreement and the court had no objective criteria against which to measure whether it was reasonable or unreasonable for a party to refuse to agree to any particular term.

Price.

2–023 [Add a new sentence to the end of note 114: page 35]

In *MRI Trading AG v Erdenet Mining Corporation LLC* [2013] EWCA Civ 156, CA where a shipping schedule and certain terms were to "be agreed during the negotiations of terms for 2010 terms", it was held that the terms were enforceable.

Contract to negotiate.

[Substitute note 125: page 37]　　　　　　　　　　　　　　　　　　　　　2–024

[125] In *Sulamerica Cia Nacional de Seguros SA v Enesa Engenharia SA* [2012] EWCA Civ 638, [2013] 1 W.L.R. 102 the Court of Appeal held that the mediation clause was incapable of giving rise to a binding obligation of any kind but referred with approval to the decisions of Colman J. in *Cable & Wireless plc v IBM United Kingdom Ltd* [2002] 2 All E.R. (Comm) 1041 and Ramsey J. in *Holloway v Chancery Mead Ltd* [2008] 1 All E.R. (Comm) 653 that an agreement to enter into a prescribed procedure for mediation is capable of giving rise to a binding obligation, provided that matters essential to the process do not remain to be agreed. See also, *Wah v Grant Thornton International Ltd* [2012] EWHC 3198 (Ch).

[Substitute the final sentence of para.2–024 with the following: page 37]

It is arguable that the decision in *Walford v Miles*[126] does not decide that, in particular circumstances and within a concluded agreement, an express obligation to negotiate a particular aspect in good faith is "completely without legal substance".[127]

[126] *Walford v Miles* [1992] 2 A.C. 128, HL. Contrast the approach to the duty to negotiate in good faith in the New South Wales Supreme Court in *United Group Rail Services v Rail Corporation New South Wales* (2009) 127 Con. L.R. 202.
[127] In *Petromec Inc v Petroleo Brasileiro SA* [2006] 1 Lloyd's Rep. 121, the Court of Appeal, whilst accepting that it was bound by *Walford v Miles*, considered obiter that in the particular circumstances there were no good reasons for saying that an obligation to negotiate certain issues in good faith as to the extra cost and time of an upgrade and changes was unenforceable.

[Add new paragraph at the end of paragraph 2–024: page 37]

Agreements to use reasonable endeavours to agree (unlike agreements to use one's best endeavours to obtain a result)[127A] are unenforceable. This is due to the difficulty in drawing a line between what is to be regarded as reasonable or unreasonable in an area where the parties may legitimately have differing views or interests but have not provided for any criteria on the basis of which a third party can assess or adjudicate the matter in the event of dispute.[127B] Agreements to negotiate in good faith are unenforceable[127C] because they are unworkable as it was inherently inconsistent with the position of a negotiating party.

[127A] See *Little v Courage Ltd* [1995] C.L.C. 164, CA per Millett L.J., 169.
[127B] *Phillips Petroleum v Enron Europe* [1997] C.L.C. 329 per Potter L.J., 343; *Multiplex Constructions UK Limited v Cleveland Bridge UK Limited* [2006]

EWHC 1341, [633-639]; and *Shaker v Vistajet Group Holdings SA* [2012] EWHC 1329, [7].

[127C] In *Barbudev v Eurocom Cable Management Bulgaria Eood* [2012] EWCA Civ 548, [43]–[46] it was held that an offer to invest on terms to be agreed in an investment agreement negotiated in good faith was unenforceable.

(h) Acceptance by conduct

2–027 [Substitute note 138: page 39]

[138] *Felthouse v Bindley* (1862) 11 C.B. (N.S.) 869; *Allied Marine Transport Ltd v Vale do Rio Doce Navegacao SA (The Leonidas D)* [1985] 1 W.L.R. 925 per Robert Goff L.J., at 937E: "It is well settled that that principle [of equitable estoppel] requires that one party should have made an unequivocal representation that he does not intend to enforce his strict legal rights against the other; yet it is difficult to imagine how silence and inaction can be anything but equivocal"; and *Liberty Insurance Pte Ltd v Argo Systems FZE* [2011] EWCA Civ 1572, [46].

[Substitute note 139: page 39]

[139] *Vitol S.A. v Norelf Ltd* [1996] A.C. 800, 812.

(i) Acceptance by post, telex, fax or email

2–028 [Substitute last sentence of second paragraph of 2–028: page 39]

Where a method of acceptance is stipulated in an express term, it will require clear words to exclude other methods of acceptance.[146A] Where other methods of acceptance are permissible an acceptance which accomplishes the object as well as or better than the stipulated method may be effective acceptance.[147]

[146A] *Yates Building v Pulleyn* (1975) 237 E.G. 183, CA (form of posting directory rather than mandatory, permissive, or obligatory); *Manchester Diocesan Council for Education* [1970] 1 W.L.R. 242; *Edmund Murray Ltd v B.S.P. International Foundations Ltd* (1992) 33 Con. L.R. 1, CA; *Anglian Water Services v Laing O'Rourke* [2010] EWHC 1529 (TCC), [34–50]; *Ener-G Holdings Plc v Hornell* [2012] EWCA Civ 1059, CA. (Lord Neuberger M.R. and Gross L.J., Longmore L.J. dissenting).
[147] *Tinn v Hoffman & Co* (1873) 29 L.T. 271, 274; *Manchester Diocesan Council for Education* [1970] 1 W.L.R. 242.

OFFER AND ACCEPTANCE

Contracts formed by estoppel.

[Add to text at end of paragraph 2–031: page 42]   2–031

Estoppel by convention has also been argued unsuccessfully in relation to the identity of the parties to a construction contract. In *Liberty Mercian Ltd v Cuddy Civil Engineering Ltd*[166A] the party carrying out the works was not a party to the contract but the contractual mechanism had been operated as though it were. Despite this, the court held that both parties knew that the contracting party and the party carrying out the works were distinct entities and so there was no common assumption that the party carrying out the works was the same party as the party to the contract.

[166A] (2013) 150 Con. L.R. 124; [2013] EWHC 4110 (TCC).

Entire agreement.

[Add to text at end of note 167: page 42]   2–032

Mileform Ltd v Interserve Security Ltd [2013] EWHC 3386 (QB).

[Add to text at end of paragraph 2–032: page 42]

An entire agreement clause may not preclude reliance upon representations that could ground an estoppel.[168A]

[168A] *Mears Ltd v Shoreline Housing Partnership Ltd* [2013] C.P. Rep. 39; [2013] EWCA Civ 639.

3. FORMALITIES OF CONTRACT

(a) Contracts requiring writing

Contracts for the sale of land.

[Substitute third sentence of first paragraph: page 43]   2–034

An option to purchase land is such a contract but a notice exercising an option to purchase is not,[172] neither is an independent collateral contract,[173] nor a "lock-out" agreement relating to the sale of land,[174] nor any other agreement that has a disposing effect but which lacks a disposing purpose.[174A]

[174A] *Yeates v Line* [2013] Ch. 363, [2012] EWHC 3085 (Ch), following the ruling in *Joyce v Rigolli* [2004] EWCA Civ 79.

Contracts of suretyship and guarantee

2–035 [Add to end of note 188: page 45]

Section 4 was intended to prevent the court from having to resolve disputes as to oral utterances not to prevent it considering a contract of guarantee identified from a sequence of negotiating emails or other documents which could constitute an agreement in writing: *Golden Ocean Group Ltd v Salgaocar Mining Industries Pvt Ltd* [2012] EWCA Civ 265, [21]–[22].

[Amend reference in note 192]

see *Chitty on Contracts*, edited by H. Beale, 31st edn (London: Sweet & Maxwell, 2012), Vol.2, paras 44–008, 44–043 and 44–046.

4. CAPACITY OF PARTIES

(b) Aliens

2–039 [Amend reference in note 207: page 47]

see further, *Chitty on Contracts*, edited by H. Beale, 31st edn (London: Sweet & Maxwell, 2012), Vol.1, para.11–024.

(e) Corporations

2–042 [Substitute note 213: page 48]

See *Chitty on Contracts*, edited by H. Beale, 31st edn (London: Sweet & Maxwell, 2012), Vol.1, paras 9–001 and 9–002.

(i) Unincorporated associations

2–047 [Substitute note 224: page 50]

See further, *Chitty on Contracts*, edited by H. Beale, 31st edn (London: Sweet & Maxwell, 2012), Vol.1, para.9–072 and following; for trade unions, see para.9–081 and following.

Chapter Three

CONSTRUCTION OF CONTRACTS

■ A.	**Construction of contracts**	3–001
☐ 1.	Expressed intention	3–002
	☐ (a) *Extrinsic evidence—not normally admissible*	3–003
	☐ (b) *Extrinsic evidence—when admissible*	3–009
☐ 2.	Rules of construction	3–021
3.	Alterations	3–038
☐ 4.	Implied terms	3–039
	☐ (a) *Statutory implication*	3–040
	☐ (b) *Necessary implication*	3–042
☐ 5.	Construction of deeds	3–061
6.	The value of previous decisions	3–062
☐ B.	**Risk, indemnity and exclusion clauses**	3–064
1.	Risk clauses	3–067
☐ 2.	Indemnity clauses	3–068
☐ 3.	Exclusion clauses	3–072
■ 4.	Legislation affecting exclusion and limitation clauses	3–074
☐ 5.	Unfair Contract Terms Act 1977	3–075

A. Construction of Contracts

[Amend references in note 1: page 51] **3–001**

K. Lewison, *Interpretation of Contracts*, 5th edn (London: Sweet & Maxwell, 2011).
Chitty on Contracts, edited by H.G. Beale, 31st edn (London: Sweet & Maxwell, 2012), para.12–041 et seq.

CHAPTER THREE—CONSTRUCTION OF CONTRACTS

1. EXPRESSED INTENTION

3–002 [Add at end of note 5: page 52]

See also, *Rainy Sky S.A. v Kookmin Bank* [2011] UKSC 50, [2011] 1 W.L.R. 2900, [14] and [21]; *Berrisford (FC) v Mexfield Housing Co-Operative Ltd* [2011] UKSC 52, [2012] A.C. 955, [17] and [107]; *Lloyds TSB Foundation for Scotland v Lloyds Banking Group plc (Scotland)* [2013] UKSC 3, [2013] 1 W.L.R. 366, [43]. However, application of "commercial common sense" is not without limits; see *Aston Hill Financial Inc v African Minerals Finance Ltd* [2013] EWCA Civ 416, approving the following statement in *Jackson v Dear* [2012] EWHC 2060 (Ch), per Briggs J. at [40]: "The dictates of common sense may enable the court to choose between alternative interpretations (with or without implied terms), not merely where one would 'flout' it, but where one makes more common sense than the other. But this does not elevate commercial common sense into an overriding criterion, still less does it subject the parties to the individual judge's own notions of what might have been the most sensible solution to the parties' conundrum."

(a) Extrinsic evidence—not normally admissible

Blanks.

3–004 [Add at end of note 12: page 53]

Absent extrinsic evidence, it may be possible to fill the blank on the basis of the remaining terms of the contract and the otherwise admissible factual matrix: see *Westvilla Properties Ltd v Dow Properties Ltd* [2010] 2 P. & C.R. 19; [2010] EWHC 30 (Ch).

Deletions from printed documents.

3–007 [Amend references in note 23: page 55]

Chitty on Contracts, edited by H.G. Beale, 31st edn (London: Sweet & Maxwell, 2012), para.12-069, fn.307.

[Add at end of note 24: page 55]

See also, *Ted Baker Plc v AXA Insurance UK Plc* [2012] EWHC 1406 (Comm), [84].

EXPRESSED INTENTION

(b) Extrinsic evidence—when admissible

Surrounding circumstances.

[Add at end of note 45: page 57]   3–011

See also the observations in *Scottish Widows Fund and Life Assurance Society v BGC International (formerly Cantor Fitzgerald International)* [2012] EWCA Civ 607, [34–35], on the difficulties of using negotiations as evidence of the general object of the transaction.

Custom and usage.

[Add to text at end of paragraph: page 59]   3–013

In circumstances where trade practice is insufficiently notorious to amount to a custom, that practice may be relevant as part of the factual context known to both parties.[54A]

[54A] *Crema v Cenkos Securities plc* [2011] 1 W.L.R. 2066; [2010] EWCA Civ 1444.

To show collateral warranties.

[Add at end of note 69: page 61]   3–018

Thinc Group v Armstrong [2012] EWCA Civ 1227, [85]–[87].

[Add at end of note 70: page 61]

For a case where a collateral warranty was held to exist, see *Thinc Group v Armstrong* [2012] EWCA Civ 1227, [89]–[91].

2. RULES OF CONSTRUCTION

General rules of construction.

[Add at end of note 84: page 63]   3–022

For a case where the Court of Appeal rejected an attempt to derive a mistake from pre-contractual negotiations, see *Scottish Widows Fund and Life Assurance Society v BGC International (formerly Cantor Fitzgerald International)* [2012] EWCA Civ 607, [24] and [69–70].

[Add at end of note 87: page 63]

See also, *Fitzroy House Epworth Street (No.1) Ltd v Financial Times Ltd* [2006] EWCA Civ 329, [2006] 1 W.L.R. 2207 in which Sir Andrew Morritt C. referred to *Brutus v Cozens* saying that: "The application of an ordinary

English word to a set of primary facts is itself a question of fact"; *Giles v Tarry* [2012] EWCA Civ 837, [60–61].

3–023 [Add at end of note 96: page 64]

There must be a clear mistake or something commercially nonsensical for the court to depart from the ordinary meaning: see *Campbell v Daejan Properties Ltd* [2012] EWCA Civ 1503, [64] and [72].

Correction of mistakes by construction.

3–024 [Add at end of note 98: page 64]

See, for an example in the context of a construction project, *Liberty Mercian Ltd v Cuddy Civil Engineering Ltd* (2013) 150 Con. L.R. 124; [2013] EWHC 4110 (TCC).

[Add at end of note 99: page 64]

For correction of mistakes by construction, see also the cases cited at fnn.77 and 84, above.

Clerical errors corrected.

3–028 [Add at end of note 105: page 65]

For a consideration of the meaning of "clerical error", in the context of a will, see *Marley v Rawlings* [2014] 2 W.L.R. 213; [2014] UKSC 2.

Contra proferentem rule.

3–032 [Add at end of note 125: page 68]

"A court should be wary of starting its analysis by finding an ambiguity by reference to the words in question looked at on their own. And it should not, in any event, on such a finding, move straight to the contra proferentem rule without first looking at the context and, where permissible, aids to identifying the purpose of the commercial document of which the words form part": *Direct Travel Insurance v McGeown* [2004] 1 All E.R. (Comm) 609; [2003] EWCA Civ 1606, per Auld L.J., at [13].

Recitals.

3–033 [Substitute last sentence of note 135: page 69]

For misstatement of facts in recitals in deeds, see para.3–061.

Irreconcilable clauses.

[Amend note 137: page 69]: 3–034

Forbes v Git [1922] 1 A.C. 256, 259, PC referred to in *Westcoast (Holdings) Ltd v Wharf Land Subsidiary (No.1) Ltd* [2012] EWCA Civ 1003, [61]. See also, "Meaningless words" at para.2–022.

Interpretation of contracts made with a public authority.

[Substitute note 144: page 70] 3–037

For a fuller discussion of these matters see *Chitty on Contracts*, edited by H.G. Beale, 31st edn (London: Sweet & Maxwell, 2012), para.1–076.

4. IMPLIED TERMS

[Amend reference in note 151] 3–039

For an interesting discussion of terminology, see *Halsbury's Laws of England*, edited by Lord Mackay of Clashfern, 5th edn (London: LexisNexis, 2014), Vol.22, para.364 and following.

(a) Statutory implication

Housing Grants, Construction and Regeneration Act 1996.

[Add at end of note 158] 3–040

There is currently uncertainty whether, in a contract incorporating an adjudication provision, there is a further implied that an unsuccessful party to an adjudication who is successful in a final determination by litigation is entitled to recover any payment made in compliance with the adjudicator's decision: see *Aspect Contracts (Asbestos) Ltd v Higgins Construction plc* [2014] 1 W.L.R. 1220; [2013] EWCA Civ 1541 (currently the subject of an appeal to the Supreme Court); cf. *Walker Construction (UK) Ltd v Quayside Homes Ltd* (2014) 153 Con. L.R. 26, [2014] EWCA Civ 93. See also P. Sheridan, "Adjudication and limitation: implied term and restitution", Const. L.J. 2014, 30(1), 57–66.

Supply of Goods and Services Act 1982.

[Add note to text at end of first paragraph: page 72] 3–041

[... implies in a contract for services.][169A]

[169A] For an example, relating to a "Group Repair Scheme" under Chapter II (now repealed) of the Housing Grants, Construction and Regeneration Act

1996, see *Cometson v Merthyr Tydfil County BC* [2012] 50 E.G. 101 (C.S.); [2012] EWHC 3446 (Ch).

(b) Necessary implication

3–042 [Add at end of note 170: page 73]

The basis in *Stirling v Maitland* has no application to circumstances where one party prevents a condition precedent to the other party's performance being fulfilled (i.e. where an alteration of the assumed state of affairs has been prevented): *Swallowfalls Ltd v Monaco Yachting and Technologies SAM* [2014] 2 All E.R. (Comm) 185; [2014] EWCA Civ 186, [34].

(i) Implication to make contract work

3–043 [Add to end of note 176: page 74]

Greater London Council v Cleveland Bridge and Engineering Co Ltd was followed and applied in *Leander Construction Ltd v Mullalley & Co Ltd* [2011] EWHC 3449 (TCC), [32]–[44]. Where the contract imposes a unilateral obligation without specifying the time in which it is to be fulfilled, "the parties cannot have intended the obligation to be of perpetual or indefinite duration": *Wuhan Ocean Economic & Technical Cooperation Co Ltd v Schiffahrts-Gesellschaft Hansa Murcia mbH & Co KG* [2013] 1 All E.R (Comm) 1277; [2012] EWHC 3104 (Comm), [24]. On the implication of a term requiring good faith in the exercise of discretion under a contract, see *Mid Essex Hospital v Compass Group UK and Ireland Ltd* [2013] B.L.R. 265; [2013] EWCA Civ 200.

(ii) Implication of "usual" terms—employer

3–046 [Add at end of note 200: page 77]

See also, in the context of a shipbuilding contract, the court's decision on the cross-appeal in *Swallowfalls Ltd v Monaco Yachting and Technologies SAM* [2014] 2 All E.R. (Comm) 185; [2014] EWCA Civ 186.

(iii) Implication of "usual" terms—contractor

Warranties of fitness for purpose and good quality.

3–052 [Add at end of final paragraph: page 83]

No warranty of quality can be implied where the quality of the thing supplied cannot be assessed in the abstract. For example, where the thing supplied is a system, its quality can only be assessed by reference to its intended or actual application, because "what may be a good system for one application may not be a good system for another application".[244A]

IMPLIED TERMS

[244A] *Trebor Bassett Holdings Ltd v ADT Fire & Security Plc* (2012) 144 Con. L.R. 1; [2012] EWCA Civ 1158, per Tomlinson L.J., at [46].

Exclusion of warranties.

[Substitute note 247: page 83]   3–053

See cases cited in fn.235, above.

[Add at end of note 249: page 84]

For an example of reliance see *BSS Group PLC v Makers (UK) Ltd* [2011] EWCA Civ 809, [43]–[45].

Fitness for purpose of completed works.

[substitute note 267: page 85]   3–054

[267] *Steel Co of Canada Ltd v Willand Management Ltd* [1966] S.C.R. 746 (Canada SC); *Greater Vancouver Water District v North American Pipe & Steel Ltd* (2012) BCCA 337 (British Columbia CA) reversing the decision at (2011) BSCS 30.

Sale of buildings.

[Add at end of note 278: page 87]   3–058

See also *Harrison v Shepherd Homes Ltd* (2011) 27 Const. L.J. 709; [2011] EWHC 1811 (TCC).

5. CONSTRUCTION OF DEEDS

[Substitute note 295: page 89]   3–061

Greer v Kettle [1938] A.C. 156, 171; *Prime Sight Ltd v Lavarello* [2014] A.C. 436, [2013] UKPC 22, and the cases cited therein; see also generally *Halsbury's Laws of England*, edited by Lord MacKay of Clashfern, 5th edn (London: LexisNexis, 2014), Vol.32, para.257.

B. RISK, INDEMNITY AND EXCLUSION CLAUSES

Loss caused by negligence.

[Add at end of note 306: page 90]   3–065

In a chain of building contracts "it is not inherently unlikely that each party will agree to be liable for shortcomings in its own work, even if superior parties in the chain fail to detect those shortcomings", even where such

CHAPTER THREE—CONSTRUCTION OF CONTRACTS

failure is negligent: *Greenwich Millennium Village Ltd v Essex Services Group Plc (formerly Essex Electrical Group Ltd)* [2014] EWCA Civ 960, per Jackson L.J., at [92].

[Add to text after fifth sentence in second paragraph: "But if there is no ... losses caused by negligence": page 91]

The principles stated above should not be applied mechanistically and ought to be regarded as no more than guidelines. They do not provide an automatic solution to any particular case. The court's function is always to interpret the particular contract in the context in which it was made.[310A]

[310A] *HIH Casualty & General Insurance Ltd v Chase Manhattan Bank* [2003] UKHL 6, [11], [61–63], [95] and [116]; *Mir Steel UK Ltd v Christopher Morris* [2012] EWCA Civ 1397, [31–35]; *Greenwich Millennium Village Ltd v Essex Services Group Plc (formerly Essex Electrical Group Ltd)* [2014] EWCA Civ 960, per Jackson L.J., at [94].

2. INDEMNITY CLAUSES

Limitation period for liability under indemnity clauses.

3–070 [Amend reference in note 339: page 94]

... cases cited in *Halsbury's Laws of England*, edited by Lord MacKay of Clashfern, 5th edn (London: LexisNexis, 2014), Vol.68, para.967;

3. EXCLUSION CLAUSES

Third parties.

3–073 [Add new paragraph at end of 3–073: page 97]

There are also circumstances (common to construction projects) in which an exclusion clause can be relied upon against a third party, namely where that third party and the beneficiary of the exclusion clause cause a loss to the benefactor for which they are jointly and severally liable. The exclusion clause, assuming it excludes the beneficiary's liability for the loss suffered by the benefactor, can be relied on by the beneficiary as a defence to any claim brought against it by the third party under the Civil Liability (Contribution) Act 1978.[361A] Where such a defence is successfully argued, the third party will be liable for the whole of the loss for which it is jointly responsible:

"[N]o wrongdoer has a right to assume that there will be other wrongdoers available to contribute to the liability which he incurs; and there are also many reasons, legal and factual, why any expectation which he may unwisely hold to that effect may be frustrated."[361B]

[361A] *Farstad Supply AS v Enviroco Ltd* [2010] 2 Lloyd's Rep. 387; [2010] UKSC 18, dealing with analogous provisions under the Law Reform (Miscellaneous Provisions) (Scotland) Act 1940.
[361B] See *Farstad Supply AS v Enviroco Ltd* [2010] 2 Lloyd's Rep. 387; [2010] UKSC 18, per Lord Mance, at [54].

Injunctive relief.

[Add new paragraph 3–073A: page 97]   3–073A

Where one party to a contract stipulates that its liability to another party, in the event that it breaches its obligations under that contract, is to be limited or excluded by the operation of an exclusion clause, that is a factor likely to support the grant of an injunction to prohibit any breach in the first place.[361C] That the clause in question was agreed will not be sufficient to defeat the innocent party's application.[361D]

[361C] *AB v CD* [2014] 3 All E.R. 667; [2014] EWCA Civ 229.
[361D] See *AB v CD* [2014] 3 All E.R. 667; [2014] EWCA Civ 229, per Underhill L.J., at [30].

4. LEGISLATION AFFECTING EXCLUSION AND LIMITATION CLAUSES

[Amend reference in note 365: page 98]   3–074

see *Chitty on Contracts*, edited by H.G. Beale, 31st edn (London: Sweet & Maxwell, 2012), Ch.15.

5. UNFAIR CONTRACT TERMS ACT 1977

[Add at end of note 369: page 99]   3–075

See also, *Kingspan Environmental Ltd v Borealis* [2012] EWHC 1147 (Comm), [569]–[587].

Liability arising in contract.

[Substitute final sentence of note 388: page 101]   3–078

For cases where the requirement of reasonableness was satisfied, see *Chester Grosvenor v Alfred McAlpine* (1991) 56 B.L.R. 115; and *West v Ian Finlay and Associates* (2014) 153 Con. L.R. 1, [2014] EWCA Civ 316.

[Substitute final sentence of second paragraph: page 101]

Thus a clause in written standard terms of business seeking to exclude rights of set-off in wide terms may fail to satisfy the requirement of reasonableness and be held to be ineffective.[390]

[Add at end of note 390: page 101]

But see *F.G. Wilson (Engineering) Ltd v John Holt & Co (Liverpool) Ltd* [2012] EWHC 2477 (Comm), [93]–[109] where it was held that on the facts, a clause in the claimant's standard terms and conditions which provided "Buyer shall not apply any set-off to the price of Seller's products without prior written agreement by the Seller" satisfied the requirement of reasonableness.

Chapter Four

THE RIGHT TO PAYMENT AND VARIED WORK

A.	**The right to payment**	4–001
1.	Lump-sum contracts	4–002
	☐ (a) *Entire contracts*	4–003
	☐ (b) *Substantial performance*	4–008
	☐ (c) *Non-completion*	4–010
2.	Contracts other than for a lump sum	4–016
☐ 3.	Quantum meruit	4–020
B.	**Varied work**	4–023
1.	What is extra work?	4–024
	☐ (a) *Lump-sum contract for whole work—widely defined*	4–025
	☐ (b) *Lump-sum contract for whole work—exactly defined*	4–027
	☐ (c) *Measurement and value contracts*	4–030
	☐ (d) *Other common features*	4–033
2.	Agent's authority	4–038
☐ 3.	Payment for extra work	4–040
☐ 4.	Rate of payment	4–053
5.	Appropriation of payments	4–056

A. The Right to Payment

1. LUMP-SUM CONTRACTS

(a) Entire contracts

[Add at end of note 11: page 106]   **4–003**

The appointment of an adjudicator involved performance in the form of an enforceable decision and there was nothing in the contract to indicate that the parties agreed that they would pay for an unenforceable decision or that they would pay for the services performed by the adjudicator which were

CHAPTER FOUR—THE RIGHT TO PAYMENT AND VARIED WORK

preparatory to the making of an unenforceable decision. A decision which was unenforceable was of no value to the parties: see *P.C. Harrington Contractors Ltd v Systech International Limited* [2012] EWCA Civ 1371, [32].

[Add at end of note 12: page 106]

See also, *Smales v Lea* [2011] EWCA Civ 1325, [43]:

"In modern construction of contracts and contracts of retainer for professional services, it is relatively unusual for the client to have no obligation to make any payment unless and until the contractor or the professional firm has performed every single one of the obligations undertaken."

Retention money.

4–005 [Add new paragraph at end of text: page 109]

Where it transpires that a condition precedent to release of the retention money will never be met and the contract provides no mechanism for what is to be done in those circumstances, the court may take a purposive approach to discerning the intention of the parties. As the usual purpose of retention is to provide the employer with security against the possibility that the contractor might fail to perform its obligations, this is likely to result in the retention money becoming payable to the contractor as the outstanding part of the purchase price, subject to a set-off (where applicable) in respect of any losses suffered by the employer as a result of the contractor's breach of contract.[35A]

[35A] *Relicpride Building Co Ltd v Cordara* (2013) 147 Con. L.R. 92; [2013] EWCA Civ 158.

Recovery of money paid.

4–007 [Amend reference in note 47: page 110]

Lord Goff of Chieveley and G. Jones, *Goff and Jones, The Law of Unjust Enrichment*, 8th edn (London: Sweet & Maxwell, 2011), para.3–035 and following.

(b) Substantial performance

4–008 [Substitute last sentence of note 55: page 112]

For sale of goods cases, cf. *Regent v Francesco of Jermyn St* [1981] 3 All E.R. 327 and *Foxholes Nursing Home Ltd v Accora Ltd* [2013] EWHC 3712 (Ch).

(c) Non-completion

Acceptance.

[Substitute note 73: page 114] 4–012

Lord Goff of Chieveley and G. Jones, *Goff and Jones, The Law of Unjust Enrichment*, 8th edn (London: Sweet & Maxwell, 2011).

Unpaid instalments.

[Add at end of note 85: page 115] 4–014

See also, *Cavenagh v William Evans Limited* [2012] EWCA Civ 697, [41] which adopted [44] of the judgment of Moore-Bick L.J. in *Stocznia Gdynia SA v Gearbulk Holdings Ltd*.

3. QUANTUM MERUIT

[Add new sentence and quotation after third paragraph in text ("However, more recently ... pay a reasonable sum.[101]"): page 118] 4–020

The Supreme Court, in dealing with a claim for unjust enrichment, drew the following distinction between two types of quantum meruit as follows:[101A]

> "It is common ground that the correct approach to the amount to be paid by way of a quantum meruit where there is no valid and subsisting contract between the parties is to ask whether the defendant has been unjustly enriched and, if so, to what extent. The position is different if there is a contract between the parties. Thus, if A consults, say, a private doctor or a lawyer for advice there will ordinarily be a contract between them. Often the amount of his or her remuneration is not spelled out. In those circumstances, assuming there is a contract at all, the law will normally imply a term into the agreement that the remuneration will be reasonable in all the circumstances. A claim for such remuneration has sometimes been referred to as a claim for a quantum meruit. In such a case, while it is no doubt relevant to have regard to the benefit to the defendant, the focus is not on the benefit to the defendant in the way in which it is where there is no such contract."

[101A] *Benedetti v Sawiris* [2013] UKSC 50, [2013] 3 W.L.R. 351 per Lord Clarke of Stone-cum-Ebony, [9].

[Add at end of note 103: page 118]

This may also be true where the agreed price is conditional and that condition has not been fulfilled: see *Tallington Lakes Ltd (No.01595671) v*

Larking Gowen (A Firm) [2014] EWCA Civ 959, a case involving accountancy services.

[Amend references in note 106: page 118]

Lord Goff of Chieveley and G. Jones, *Goff & Jones, The Law of Unjust Enrichment*, 8th edn (London: Sweet & Maxwell, 2011), Ch.16; *Chitty on Contracts*, edited by H.G. Beale, 31st edn (London, Sweet & Maxwell, 2012), Vol.1, para.29–076.

[Amend reference in note 109: page 119]

Lord Goff of Chieveley and G. Jones, *Goff & Jones, The Law of Unjust Enrichment*, 8th edn (London: Sweet & Maxwell, 2011), Ch.16.

Assessment of a reasonable sum.

4–021 [Add new note 121A at end of first paragraph, second sentence (following "framework"): page 121]

[121A] See *Benedetti v Sawiris* [2013] UKSC 50, [2013] 3 W.L.R. 351 per Lord Clarke of Stone-cum-Ebony at [9], quoted at para.4–020, above.

[Delete last two sentences of para.4–021 and replace with the following new sentences. Also substitute notes 134 and 135 and add new note 135A: page 122]

In a claim for quantum meruit, statutory interest is a matter for the exercise of the court's discretion and may not run from the generally applicable date of the accrual of the cause of action.[134] In a commercial setting, it would be proper to take account of the manner and the time at which persons acting honestly and reasonably would pay.[135] Therefore, it may run from the date when the sum due was ascertainable which was when the Contractor had furnished its final account and the Employer had been given a reasonable opportunity to consider it.[135A]

[134] Goff J. in *BP Exploration Co (Libya) Ltd v Hunt* [1979] 1 W.L.R. 783.
[135] *General Tire & Rubber Co v Firestone Tire & Rubber Co Ltd* [1975] 1 W.L.R. 819, CA per Lord Wilberforce, at 836; *Sycamore Bidco Ltd v Breslin* [2013] EWHC 174 (Ch), [12].
[135A] *Claymore Services Ltd v Nautilus Properties Ltd* [2007] B.L.R. 452; *Sycamore Bidco Ltd v Breslin* [2013] EWHC 174 (Ch), [11–12].

B. Varied Work

1. WHAT IS EXTRA WORK?

(a) Lump-sum contract for whole work—widely defined

Indispensably necessary works.

[Substitute (iii): page 125] 4–025

(iii) Unexpected labour caused by difficulties of the terrain,[148] by the proposed method of carrying out the works,[149] or by greater than anticipated scope of the work.[149A]

[149A] *Atkins Ltd v Secretary of State for Transport* (2013) 146 Con. L.R. 169; [2013] EWHC 139 (TCC).

(b) Lump-sum contract for whole work—exactly defined

Bills of quantities contract.

[Amend reference in note 159: page 126] 4–027

see para.4–031.

(c) Measurement and value contracts

[Amend reference in note 177: page 128] 4–030

see para.4–017.

(d) Other common features

Omissions.

[Delete final four sentences of the paragraph and also notes 190 and 191: page 130] 4–033

3. PAYMENT FOR EXTRA WORK

Emergency.

[Substitute note 220: page 134] 4–042

Lord Goff of Chieveley and G. Jones, *Goff & Jones, The Law of Unjust Enrichment,* 8th edn (London: Sweet & Maxwell, 2011), para.18–50 and following.

CHAPTER FOUR—THE RIGHT TO PAYMENT AND VARIED WORK

Architect's final certificate.

4–052 [Add at end of note 268: page 140]

For a case which raised the issue of when a challenge to an architect's certificate had commenced for the purpose of the saving proviso in a conclusive evidence clause, see *Brighton University v Dovehouse Interiors Ltd* (2014) 153 Con. L.R. 147; [2014] EWHC 940 (TCC).

4. RATE OF PAYMENT

Work within the contract.

4–053 [Add at end of paragraph: page 140]

Where the contract does not specify a rate but the contractor later proposes a rate to which the employer agrees, the employer cannot subsequently challenge that rate on the basis that it is unreasonable:

"If the employer agrees to pay that rate for the work, then that becomes, as between the employer and the contractor, the reasonable rate for that particular work."[272A]

[272A] *Oakrock Ltd v Travelodge Hotels Ltd* [2014] EWHC 1868 (TCC), [22].

Chapter Five

EMPLOYER'S APPROVAL AND ARCHITECT'S CERTIFICATES

A. Employer's approval	5–001
☐ B. Architect's certificates	5–007
1. Types of certificate	5–007
☐ 2. Certificates as condition precedent	5–014
☐ 3. Recovery of payment without certificate	5–017
☐ 4. Binding and conclusive certificates	5–023
5. Attacking a certificate	5–029
(a) *Not within the architect's jurisdiction*	5–030
☐ (b) *Not properly made*	5–034
(c) *Disqualification of the certifier*	5–040
(d) *Effect of an arbitration clause*	5–045

B. Architect's Certificates

1. TYPES OF CERTIFICATE

Retention money.

[Add to note 67: page 151]   **5–013**

See also para.4–005 above.

2. CERTIFICATES AS CONDITION PRECEDENT

[Add at end of note 75: page 152]   **5–014**

See also *Laing O'Rourke Construction Ltd (formerly Laing O'Rourke Northern Ltd) v Healthcare Support (Newcastle) Ltd* [2014] EWHC 2595 (TCC), where it was held that defects that had no "material adverse impact"

CHAPTER FIVE—EMPLOYER'S APPROVAL AND ARCHITECT'S CERTIFICATES

on the employer's enjoyment and use of the buildings ought not prevent the issue of a final certificate.

3. RECOVERY OF PAYMENT WITHOUT CERTIFICATE

Prevention by the employer.

5–020 [Add at the end of note 101: page 155]

In *R&C Electrical Engineers Ltd v Shaylor Construction Ltd* [2012] B.L.R. 373, [61] the court distinguished between circumstances which prevented the contractual machinery being operated, and circumstances in which one party refused to operate that machinery although in a position to do so, where the problem was capable of being cured.

When is a certificate binding and conclusive?

5–024 [Substitute note 125: page 159]

Cf. *Halsbury's Laws of England*, edited by Lord MacKay of Clashfern, 5th edn (London: LexisNexis, 2014), Vol.6, para.332.

4. BINDING AND CONCLUSIVE CERTIFICATES

Expiration of time.

5–028 [Add at end of note 166: page 164]

See also, more recently, *Brighton University v Dovehouse Interiors Ltd* (2014) 153 Con. L.R. 147; [2014] EWHC 940 (TCC).

5. ATTACKING A CERTIFICATE

(b) Not properly made

Mistake by architect.

5–039 [Add at the end of note 203: page 168]

The circumstances may lead to the conclusion that the certificate had not in fact been relied upon: see for example *Hunt v Optima (Cambridge) Ltd* [2014] EWCA Civ 714, where the purchasers of several flats had not received the certificate until after completing the transaction that caused their losses.

[Add at the end of note 205: page 169]

A departure from substantive instructions will be material and automatically invalidate a decision unless it is trivial or de minimis. However, a departure from express or implied procedural instructions or an unfairness will not always do so: *Ackerman v Ackerman* [2011] EWHC 3428 (Ch), [274].

CHAPTER SIX

EXCUSES FOR NON-PERFORMANCE

■ 1.	Inaccurate statements generally	6–001
☐ 2.	Misrepresentation—before 1964	6–006
3.	Negligent misstatement	6–013
☐ 4.	Misrepresentation Act 1967	6–014
☐ 5.	Collateral warranty	6–027
☐ 6.	Death	6–030
☐ 7.	Frustration and impossibility	6–032
☐ 8.	Illegality	6–048
9.	Economic duress	6–058
10.	Default of other party	6–060
	☐ (a) *Repudiation generally*	6–060
	☐ (b) *Repudiation by contractor*	6–074
	(c) *Repudiation by employer*	6–080
	(d) *Party cannot rely on own wrong*	6–090

1. INACCURATE STATEMENTS GENERALLY

Misrepresentation.

[Add to the end of note 4: page 176]   6–003

A representation is treated as true if it is *"substantially correct"*: the difference between what is represented and what is in fact the case would not be considered material: *Avon Insurance plc v Swire Fraser Ltd* [2000] 1 All E.R. (Comm) 573. The representation need not be entirely correct, provided it is substantially correct. The difference would not have been likely to induce a reasonable person in its position to enter into the contract: *Raiffesen Zentralbank Osterreich AG v Royal Bank of Scotland Plc* [2010] EWHC 1392 (Comm) [2011] 1 Lloyd's Rep. 123, [149].

2. MISREPRESENTATION—BEFORE 1964

Innocent misrepresentation.

6–007 [Substitute note 24: page 178]

Lord Goff of Chievely and G. Jones, *Goff and Jones, The Law of Restitution*, 8th edn (London: Sweet & Maxwell, 2011), para.40–006.

Remedies for fraud.

6–012 [Amend reference in note 63: page 183]

Chitty on Contracts, edited by H. Beale, 31st edn (London: Sweet & Maxwell, 2012), Vol.1, para.6–046 and following.

4. MISREPRESENTATION ACT 1967

6–014 [Amend reference in note 69: page 184]

Chitty on Contracts, edited by H. Beale, 31st edn (London: Sweet & Maxwell, 2012), Vol.1, para.6–001 and following.

Servants or agents.

6–019 [Add to text at the end of the second paragraph: page 187]

Conversely, where a continuing representation is made to a person who subsequently becomes the agent of the party with whom the contract is concluded, the change in the identity of the prospective contracting party will not affect the continuing nature of the representation or the representor's continuing responsibility for its accuracy.[90A]

[90A] *Cramaso LLP v Ogilvie-Grant* [2014] 2 W.L.R. 317; [2014] UKSC 9.

Section 2(2).

6–020 [Amend reference in note 93: page 188]

See also, *Chitty on Contracts*, edited by H. Beale, 31st edn (London: Sweet & Maxwell, 2012), Vol.1, para.6–101 and following.

Generally.

6–024 [Add at the end of note 103: page 190]

Approved in *Springwell Navigation Corp v JP Morgan Chase Bank* [2010] EWCA Civ 1221, [181]. The same reasoning was applied to s.2 of the Unfair

Contract Terms Act 1977 in *Avrora Fine Arts Investment Ltd v Christine, Manson & Woods Ltd* [2012] EWHC 2198 (Ch), [144].

"Fair and reasonable".

[Add at the end of note 106: page 191]

6–025

For an example of the application of the clause in the context of a sale of land, see *Lloyd v Browning* [2014] 1 P. & C.R. 11; [2013] EWCA Civ 1637.

[Add at the end of note 107: page 191]

An entire agreement clause, read together with a provision that rights in respect of warranties and representations were waived, was sufficient to exclude liability: *Bikam OOD, Central Investment Group SA v Adria Cable Sarl* [2012] EWHC 621 (Comm), [45]–[47].

5. COLLATERAL WARRANTY

[Add at the end of note 111: page 192]

6–027

In the absence of such intent the alleged warranty is likely to be subject to the law on misrepresentation; see, for example, *Hunt v Optima (Cambridge) Ltd* [2014] EWCA Civ 714.

[Add at the end of note 112: page 192]

For a case where a collateral warranty was held to exist, see *Thinc Group v Armstrong* [2012] EWCA Civ 1227, [89]–[91].

6. DEATH

Personal contracts.

[Amend reference in note 135: page 195]

6–031

See *Chitty on Contracts*, edited by H. Beale, 31st edn (London: Sweet & Maxwell, 2012), Vol.1, paras 20–006 and 20–007.

7. FRUSTRATION AND IMPOSSIBILITY

Law Reform (Frustrated Contracts) Act 1943.

6–047 [Amend reference in note 219: page 205]

see *Chitty on Contracts*, edited by H. Beale, 31st edn (London: Sweet & Maxwell, 2012), Vol.1, para.23–074.

8. ILLEGALITY

No assistance from the court.

6–048 [Amend reference in note 223]

See generally, *Chitty on Contracts*, edited by H. Beale, 31st edn (London: Sweet & Maxwell, 2012), Vol.1, Ch.16.

Return of goods.

6–051 [Add to text after "plead its illegality to support its claim.": page 207]

However, the owner of the goods may rely on the illegal contract to found their claim, by way of an exception to the general prohibition just outlined, where they withdraw from the illegal transaction before it has been carried into effect.[241A] Following the decision of the Court of Appeal in *Patel v Mirza*, it appears that the owner's withdrawal need not be voluntary, and they may seek recovery of the goods by relying upon the illegal contract to establish that it has been frustrated.[241B]

[241A] *Taylor v Bowers* (1876) 1 QBD 291, CA; *Alexander v Rayson* [1936] 1 K.B. 169, CA; *Tinsley v Milligan* [1994] 1 A.C. 340, HL; *Tribe v Tribe* [1996] Ch. 107, CA.
[241B] [2014] EWCA Civ 1047.

10. DEFAULT OF OTHER PARTY

(a) Repudiation generally

6–060 [Add at end of note 286: page 213]

The effect of the breach is a question of fact, and therefore a matter for the trial judge: *Valilas v Januzaj* (2014) 154 Con. L.R. 38, 58 per Floyd L.J. The breach must continue to be fundamental at the time that the innocent party purports to accept the repudiation: *Ampurius Nu Homes Holdings Ltd v Telford Homes (Creekside) Ltd* (2013) 148 Con. L.R. 1; [2013] EWCA Civ

DEFAULT OF OTHER PARTY

577, [64] and [76]; see also *Urban I (Blonk Street) Ltd v Ayres* [2014] 1 W.L.R. 756; [2013] EWCA Civ 816.

Fundamental breach.

[Add to text at end of first paragraph: page 216] 6–066

Words or conduct that make future performance contingent upon the conduct of a third party will not necessarily evince an intention not to be bound, and will not be deemed to do so.[316A]

[316A] *Geden Operations Ltd v Dry Bulk Handy Holdings Inc (The Bulk Uruguay)* [2014] 2 All E.R. (Comm) 196; [2014] EWHC 885 (Comm), [21]–[23].

[Add at end of note 317: page 217]

Damages fall to be assessed on the basis that the party in repudiatory breach would have performed its obligations, notwithstanding the guilty party's declared intention not to perform the contract at all: *SC Compania Nationala de Transporturi Aeriene Romane Tarom SA v Jet2.com* [2014] EWCA Civ 87.

Arbitration agreements.

[Substitute note 328: page 218] 6–069

See *Bremer Vulkan v South India Shipping Corp* [1981] A.C. 909, 980, HL; *Turriff v Richards & Wallington* (1981) 18 B.L.R. 19; *John Downing v Al Tameer Establishment* [2002] B.L.R. 323; *BDMS Ltd v Rafael Advanced Systems* [2014] 1 Lloyd's Rep. 576, [2014] EWHC 451 (Comm); see also, para.17–041.

Acceptance of repudiation.

[Add at end of note 330: page 218] 6–070

There is no special rule stating that contracts of employment do not require acceptance: *Geys v Societe Generale* [2013] 1 A.C. 523.

Repudiation and contractual determination clauses.

[Amend reference in note 358: page 221] 6–072

See generally on this point, *Chitty on Contracts*, edited by H. Beale, 31st edn (London: Sweet & Maxwell, 2012), Vol.1, para.24–019.

(b) Repudiation by contractor

Delay.

6–077 [Add at end of note 376: page 224]

In *Ampurius Nu Homes Holdings Ltd v Telford Homes (Creekside) Ltd* (2013) 148 Con. L.R. 1; [2013] EWCA Civ 577, the Court of Appeal preferred the following formulation: "delay, even with its attendant uncertainties, will only become a repudiatory breach if and when the delay is so prolonged as to frustrate the contract." See also *Urban I (Blonk Street) Ltd v Ayres* [2014] 1 W.L.R. 756; [2013] EWCA Civ 816.

[Add to the end of first paragraph of 6–077, after: "In *Hill v London Borough of Camden*...had not by such conduct repudiated the contract.": page 224]

In *Baht v Masshouse Developments Ltd* [2012] 2 P.&C.R. DG3 the developer's failure to arrange for apartments to be completed will all due diligence was held to be a repudiatory breach. Their failure for several months after the contractor's administration to do anything on the site which promoted completion of the works signalled an intention not to be bound by the contracts with the purchasers.

Chapter Seven

NEGLIGENCE AND ECONOMIC LOSS

1.	Introduction	7–001
■ 2.	Liability for physical damage	7–004
3.	Liability for economic loss	7–010
4.	Specific categories of negligence	7–027
	☐ (a) *Negligent misstatement*	7–027
	(b) *Possible further categories*	7–041

2. LIABILITY FOR PHYSICAL DAMAGE

[Add at end of note 20: page 235]   **7–004**

Contrast *Chandler v Cape plc* [2012] EWCA Civ 525, [2012] 1 W.L.R. 3111 in which a company was held to owe a duty to the employees of a subsidiary company to advise on or ensure a safe system of work.

Negligent instructions.

[Add at end of note 52: page 240]   **7–009**

In *Cleightonhills v Bembridge Marine Ltd* [2012] EWHC 3449 (TCC) it was clarified that *Clay v Crump* was not authority for the proposition that a professional or contractor could not discharge his duty of care by employing a competent person to carry out his duties.

4. SPECIFIC CATEGORIES OF NEGLIGENCE

(a) Negligent misstatement

[Add at end of note 154: page 258]   **7–027**

Where the transaction that gives rise to the loss predates the statement, the claimant cannot have relied upon it in completing that transaction: see *Hunt v Optima (Cambridge) Ltd* [2014] EWCA Civ 714, where a claim for negligent misstatement failed because the claimants' purchase of the flats in question preceded the issue of the negligent certificates by several months.

[Add note 154A: page 258]

The law adopts a restrictive approach to any extension... intended by the maker of the statement to act upon it.[154A]

[154A] Depending on the facts of the case, the class of persons so intended may not be limited to the original representee(s). See the decision of the Supreme Court in *Cramaso LLP v Ogilvie-Grant* [2014] 2 W.L.R. 317; [2014] UKSC 9, per Lord Reed, at [29]: "A negligent misrepresentation is equally capable of having a continuing effect up until the time when the contract is concluded, where the person by whom the representation is made, or to whom it is addressed, becomes the agent of the person by whom the contract is concluded."

CHAPTER EIGHT

DELAY AND DISRUPTION CLAIMS

1.	Introduction	8–001
2.	Time for completion	8–003
	☐ (a) *Express completion date*	8–003
	■ (b) *Time of the essence*	8–005
	(c) *Reasonable time to complete*	8–012
3.	Extension of time	8–017
	(a) *Purpose of extension of time clauses*	8–017
	(b) *Grounds for an extension of time*	8–022
	☐ (c) *Concurrent causes of delay*	8–025
	☐ (d) *Procedure, notice provisions and conditions precedent*	8–029
4.	Delay analysis	8–044
	(a) *Different techniques*	8–044
	(b) *Base line programme*	8–052
	(c) *As-built programme*	8–053
	(d) *Critical path analysis*	8–054
	(e) *Primacy of the factual investigation*	8–056
5.	Disruption claims	8–057
	(a) *Definition*	8–057
	(b) *Causes of disruption*	8–058
	■ (c) *Proof of disruption*	8–062

2. TIME FOR COMPLETION

(a) Express completion date

Fixed date or period.

[Add a new note 4A to text: page 271]   8–003

The precise scope of each of... by reference to the terms of the contract as a whole.[4A]

41

CHAPTER EIGHT—DELAY AND DISRUPTION CLAIMS

⁴ᴬ For an example of the court's approach to the interpretation of ambiguous completion provisions, see *RWE Npower Renewables Ltd v J N Bentley Ltd* [2014] EWCA Civ 150.

Nature of the time obligation.

8–004 [Add a new note 15A at the end of the paragraph: page 272]

It was held that ... and the completion date in the contract.¹⁵ᴬ

¹⁵ᴬ Note also that in *Leander Construction Ltd v Mulalley & Co. Ltd.* [2011] EWHC 3449 (TCC), [2012] B.L.R. 152 it was held that ordinarily there will be no implied term in a building contract that the contractor should proceed regularly and diligently with the works prior to the contract completion date.

(b) Time of the essence

Notice making time of the essence.

8–009 [Add at the end of note 33: page 274]

In *Baht v Masshouse Developments Ltd* [2012] 2 P.& C.R. it was held that it was for the party alleging breach of an obligation to complete within a reasonable time to establish what a reasonable time would be, disregarding delays caused by the other party's failures.

3. EXTENSION OF TIME

(c) Concurrent causes of delay

Concurrent delay events and extension of time entitlement.

8–026 [Add to the end of note 95: page 283]

See a finding to this effect in *Walter Lilly & Co Ltd v DMW Developments Ltd* [2012] EWHC 1773 (TCC), [2012] B.L.R. 503, per Akenhead J., [370].

Apportionment.

8–027 [Add a new note 103A at the end of the first sentence of the second paragraph in 8–027: page 285]

It is suggested that ... placed too great a weight on the words "fair and reasonable".¹⁰³ᴬ

¹⁰³ᴬ See now *Walter Lilly & Co Ltd v DMW Developments Ltd* [2012] EWHC 1773 (TCC), [2012] B.L.R. 503 where it was held that: (i) under the JCT extension of time clause wording and similar forms of construction

contracts, as long as it is established that the relevant event relied upon as the basis of an extension of time claim is at least a concurrent cause of actual delay to the completion date of the works, the contractor will be entitled to an extension of time; and (ii) the approach of the Scottish Courts in this situation to permit an apportionment of the relevant period of delay between the parties so as to permit a partial award of an extension of time to the contractor probably does not reflect the law of England (see [362]–[370]).

(d) Procedure, notice provisions and conditions precedent

Condition precedent.

[Add at end of note 112: page 286]   8–030

See also *Obrascon Huarte Lain SA v Her Majesty's Attorney General for Gibraltar* [2014] B.L.R. 484; [2014] EWHC 1028 (TCC), per Akenhead J., at [312]: "I see no reason why this clause should be construed strictly against the Contractor and can see reason why it should be construed reasonably broadly, given its serious effect on what could otherwise be good claims for instance for breach of contract by the Employer."

[Add at end of note 113: page 286]

Cf. *Obrascon Huarte Lain SA v Her Majesty's Attorney General for Gibraltar* [2014] B.L.R. 484; [2014] EWHC 1028 (TCC).

Prospective and retrospective extension of time.

[Add at the end of note 134: page 290]:   8–036

In *Walter Lilly & Co Ltd v DMW Developments Ltd* [2012] EWHC 1773 (TCC), [2012] B.L.R. 503 it was held that Clause 25.3.3 of the JCT Standard Form of Building Contract required a consideration of what critically delayed the works as they went along as opposed to a purely retrospective exercise (see [362–365]).

Significance of the different approaches.

[Add new footnote 134A: page 290]   8–037

The choice between which approach...required to establish an entitlement.[134A]

[34A] But see *Walter Lilly & Co Ltd v DMW Developments Ltd* [2012] B.L.R. 503; [2012] EWHC 1773 (TCC), per Akenhead J., at [380]: "The debate about the "prospective" or "retrospective" approach to delay analysis was also sterile because both delay experts accepted that, if each approach was done correctly, they should produce the same result."

[Add new footnote 135A: page 290]

The SCL Delay and Disruption Protocol... hearing to decide on a disputed entitlement.[135A]

[135A] In *Walter Lilly & Co Ltd v DMW Developments Ltd* [2012] B.L.R. 503; [2012] EWHC 1773 (TCC), Akenhead J. expressed approval for a prospective approach, where the contractor's expert had employed "a sensible variant on the conventional approach of delay experts which was to review on a month by month basis what in each month was probably delaying overall completion. He then applied a cross check by reference to what actually happened" (see [381]).

5. DISRUPTION CLAIMS

(c) Proof of disruption

Introduction.

8–062 [Add a new note 177A: page 304]

In general the closer this evidence is to the "coal face" the more convincing it is likely to be.[177A]

[177A] See *Walter Lilly & Co Ltd v DMW Developments Ltd* [2012] EWHC 1773 (TCC), [2012] B.L.R. 503 where Akenhead J. stated [at 486(c)] that there is no set way to prove these elements and that it is open to contractors to prove them with whatever evidence will satisfy the tribunal to the requisite standard of proof.

[Add at the end of note 178: page 304]

[178] See also, *Walter Lilly & Co Ltd v DMW Developments Ltd* [2012] EWHC 1773 (TCC), [2012] B.L.R. 503 where Akenhead J. [at 490] accepted an approach that involved listing the relevant events relied upon, describing in prose form what additional or extended resources were deployed, and linking them to the causes of delay or disruption relied upon. All the additional or extended resources were costed in a detailed analysis which picked up allocations of time for staff and resources at particular times and applied to those allocations costs obtained from the contractor's computerised record keeping system.

[Add new note 180A: page 305]

It is suggested that ... if there is other convincing evidence of significant disruption.[180A]

¹⁸⁰ᴬ In *Cleveland Bridge UK Ltd v Severfield-Rowan Structures Ltd* [2012] EWHC 3652 (TCC) Akenhead J. considered that where it is satisfied on a balance of probabilities that some (more than de minimis) disruption must have occurred as a result of the contractor's breaches, what the court can and should do is to make a reasoned assessment albeit based on the minimum probably so attributable. See also the discussion in *Walter Lilly & Co Ltd v DMW Developments Ltd* [2012] EWHC 1773 (TCC), [2012] B.L.R. 503, [502-505] where Akenhead J. concluded that a contractor's claim based on apportionments was "reasonable, realistic and justifiable".

Disruption of plant, labour and materials.

[Add to the text at the end of the second paragraph in 8–063 after: "Quantifying and proving loss of productivity is a notoriously difficult task.": page 305] **8–063**

However, in *Walter Lilly & Co Ltd v DMW Developments Ltd*, Akenhead J. did not consider the "thickening" of preliminary costs to be a global or total costs claim if there was evidence to demonstrate that the contractor did apply a greater level of resource than originally planned for and that the linkage between the relevant event and the need to provide that greater resource is established.¹⁸¹ᴬ

¹⁸¹ᴬ [2012] EWHC 1773 (TCC), [2012] B.L.R. 503, [491].

[Add new note 183A: page 306]

If the original contract ... some evidence of the loss of productivity.¹⁸³ᴬ

¹⁸³ᴬ See *Walter Lilly & Co Ltd v DMW Developments Ltd* [2012] EWHC 1773 (TCC), [2012] B.L.R. 503 where Akenhead J. [at 492] accepted evidence that that the contractor's original prices, having been reviewed by a nationally known firm of quantity surveyors, were realistic, sensible and at a level which, if the events complained about had not happened, no net loss would have arisen.

[Add new note 189A: page 309] **8–066**

Nevertheless ... cost control systems which collect and track cost data for each project.¹⁸⁹ᴬ

¹⁸⁹ᴬ In *Walter Lilly & Co Ltd v DMW Developments Ltd* [2012] EWHC 1773 (TCC); [2012] B.L.R. 503 Akenhead J. [at 490] accepted costs that had been obtained from the contractor's computerised record keeping system and supported by reliable evidence from the contractor's witnesses.

CHAPTER NINE

FINANCIAL RECOVERY AND CAUSATION

☐	1.	Principles upon which damages are awarded	9–001
☐	2.	Various general considerations	9–011
	3.	Contractor's claims	9–025
		(a) Claims under the express provisions of the building contract	9–025
		☐ (b) Types of claim	9–027
		☐ (c) Contractor's claims for breach of contract	9–037
		☐ (d) Global claims	9–041
	4.	Employer's claims	9–044
		(a) Claims under the express provisions of the building contract	9–044
		☐ (b) Claims for breach of contract	9–045
☐	5.	Causation and concurrent causes	9–058

1. PRINCIPLES UPON WHICH DAMAGES ARE AWARDED

Claims under or for breach of the contract.

[Amend reference in note 4: page 309] **9–001**

See H. McGregor, *McGregor on Damages*, 19th edn (London: Sweet & Maxwell, 2014), Ch.10.

The second limb of Hadley v Baxendale.

[Add to note 29: page 313] **9–005**

See also, *John Grimes Partnership Ltd v Gubbins* [2013] EWCA Civ 37, [2013] B.L.R. 126 where the Court of Appeal confirmed that if the type or kind of loss was, at the time of contract, reasonably foreseeable by the defendant as not unlikely to result from his breach (had a breach been contemplated) then such a type or kind of loss was not too remote (see [17–19]).

CHAPTER NINE—FINANCIAL RECOVERY AND CAUSATION

9–006 [Add to note 37: page 315]

See also, *John Grimes Partnership Ltd v Gubbins* [2013] EWCA Civ 37, [2013] B.L.R. 126 for an application of the principles derived from *The Achilleas* in a claim for damages for delay against a negligent engineer.

Date of assessment.

9–009 [Add new note 60A: page 318]

Thus damages were not to be assessed as at the date of breach, ...at the date of the hearing.[60A]

[60A] See also the dictum of the court in *Ageas (UK) Ltd v Kwik-Fit (GB) Ltd* [2014] EWHC 2178 (QB), [35], applying *The Golden Victory*: "when assessing damages for breach of contract by reference to the value of a company or other property at the date of breach, whose value depends upon a future contingency, account can be taken of what is subsequently known about the outcome of the contingency as a result of events subsequent to the valuation date where that is necessary in order to give effect to the compensatory principle."

2. VARIOUS GENERAL CONSIDERATIONS

Loss of profits and loss of good will.

9–013 [Add to note 84: page 321]

See *Brit Inns Ltd v BDW Trading Ltd* [2012] EWHC 2143, 145 Con. L.R.181 for an example of the approach to the measure and proof of damages in claims for defective work and loss of profits.

Wasted expenditure.

9–014 [Add new sentence after "The burden...it would not have been so recovered is on the defendant.": page 321]

Indeed there is rebuttable presumption that the claimant would have recouped expenditure incurred, under the contract, in reliance upon the defendant's performance of the contract.[91A]

[91A] *Yam Seng Pte Ltd v International Trade Corporation Ltd* [2013] EWHC 111 (QB).

Various General Considerations

Third-party losses.

[Add at the end of note 100: page 323]   9–017

See also, *Pegasus Management Holdings SCA v Ernst & Young* [2012] EWHC 738 (Ch) applying this principle in the context of a transfer of assets between companies.

Tax.

[Substitute note 119: page 326]   9–020

See *British Transport Commission v Gourley* [1956] A.C. 185, HL. For a full discussion, see H. McGregor, *McGregor on Damages*, 19th edn (London: Sweet & Maxwell, 2014), Ch.17, and *Chitty on Contracts*, edited by H.G. Beale, 31st edn (London, Sweet & Maxwell, 2012), Vol.1, para.26–204 and following. See also, *Capita Alternative Fund Services (Guernsey Ltd) v Drivers Jonas* [2012] EWCA Civ 1417 for a recent application of the *Gourley* principle in an assessment of damages.

Foreign currency.

[Amend reference in note 121: page 327]   9–021

see H. McGregor, *McGregor on Damages*, 19th edn (London, Sweet & Maxwell, 2014), para.19–018 and following.

3. CONTRACTOR'S CLAIMS

(b) Types of claim

Loss of profit.

[Add new note 158A: page 333]   9–033

Even if, at that time ... recoverable if the loss of turnover increased the loss of the business.[158A]

[158A] In *Walter Lilly & Co Ltd v DMW Developments Ltd* [2012] EWHC 1773 (TCC), [2012] B.L.R. 503 it was held that the use of a formula, supported by relevant factual evidence of opportunities foregone, was a legitimate way of establishing this head of loss. However, in order to prove a claim for loss of profit and head office overheads caused by delay it had to be shown on the balance of probabilities that but for the delay it would have secured other work which would have contributed to such overheads and/or generated profit the loss would not have been sustained (see [540–543]).

CHAPTER NINE—FINANCIAL RECOVERY AND CAUSATION

"Hudson formula".

9–034 [Add at end of note 159: page 333]

See also *Roe Brickwork Ltd v Wates Construction Ltd* [2013] EWHC 3417 (TCC).

(c) Contractor's claims for breach of contract

Election to claim wasted expenditure

9-038 [Add at end of note 165: page 335]

See *Chitty on Contracts*, edited by H.G. Beale, 31st edn (London, Sweet & Maxwell, 2012), Vol.1, paras 26-027 and the cases cited therein.

Work partly carried out.

9–039 [Add at end of note 173: page 337]

See more recently, in the context of a solicitor's conditional fee agreement, *Howes Percival LLP v Page* [2013] EWHC 4104 (Ch), in which it was held that where the right to payment is conditional, the claimant cannot elect to claim on a quantum meruit (see [307]–[312]).

(d) Global claims

Definition.

9–041 [Add new note 177A: page 338]

In such a claim ... the result of the breaches or events relied upon.[177A]

[177A] In *Walter Lilly & Co Ltd v DMW Developments Ltd* [2012] EWHC 1773 (TCC), [2012] B.L.R. 503 it was held, albeit obiter dicta, that there was nothing in principle wrong with a global claim and that a contractor was not debarred from pursuing a global claim if it could otherwise have presented and proved its loss on a conventional basis or even where the contractor has itself created the circumstances where it is impractical or impossible to present or prove its claim on a conventional basis (see [486]).

Assessing global claims.

9–043 [Add at end of note 184: page 339]

See, as a further example of the pragmatic approach to global claims, *Bluewater Energy Services BV v Mercon Steel Structures BV* [2014] EWHC 2132 (TCC), [1350]–[1367].

4. EMPLOYER'S CLAIMS

(b) Claims for breach of contract

Offer to complete.

[Add to note 192: page 340]   9–046

See also, *Activa DPS Europe Sarl v Pressure Seal Solutions Ltd* [2012] EWCA Civ 943, [2012] Eu L.Rev. 758. The reasonableness (or otherwise) of refusing an offer to complete does not depend upon the existence of an express defects liability clause: *Woodlands Oak Ltd v Conwell* [2011] B.L.R. 365; [2011] EWCA Civ 254, per May L.J., at [24].

Defective work.

[Add to note 196: page 341]   9–047

In *Smales v Lea* [2011] EWCA Civ 1325, 140 Con. L.R. 70, it was held by the Court of Appeal that in modern construction contracts and contracts of retainer for the provision of professional services it was relatively unusual for a client to have no obligation to make any payment until every obligation had been carried out, although it was possible that such a contract could be construed in this manner.

Going slow.

[Add to note 242: page 347]   9–052

See also, *Leander Construction Ltd v Mulalley & Co. Ltd.* [2011] EWHC 3449 (TCC), [2012] B.L.R. 152.

Inconvenience, discomfort and distress.

[Add at end of note 244: page 347]   9–053

For a more recent example, in which the court endorsed the limits set out in *Axa Insurance*, see *West v Ian Finlay and Associates* (2014) 153 Con. L.R. 1; [2014] EWCA Civ 316, per Vos L.J., at [83]–[86].

Lack of inspection by architect.

[Add at end of note 256: page 349]   9–055

Thus a company responsible for the design and installation of mechanical works was entitled to be indemnified by its labour-only sub-contractor in respect of workmanship defects, even though it should have detected such defects on inspection: *Greenwich Millennium Village Ltd v Essex Services Group plc* [2014] EWCA Civ 960.

5. CAUSATION AND CONCURRENT CAUSES

Causation—claims under the contract.

9–058 [Add to note 262: page 350]

See also, *Nulty v Milton Keynes BC* [2013] EWCA 15, [2013] B.L.R. 134, where the Court of Appeal held that the civil balance of probability test meant no less and no more than the court having to be satisfied on rational and objective grounds that the case for believing that the suggested means of causation occurred was stronger than the case for not believing (at [35]).

Concurrent causes.

9–059 [Add to note 265: page 351]

This distinction, between the test of causation that applies to extension of time, loss and expense provisions in provisions using the JCT wording, was confirmed in *Walter Lilly & Co Ltd v DMW Developments Ltd* [2012] EWHC 1773 (TCC), [2012] B.L.R. 503 per Akenhead J., at [362]–[370] and [540]–[543].

The relevance of common law causation principles.

9–060 [Add to note 267: page 351]

For the unavailability of the defence of contributory negligence to a claim for breach of a strict contractual duty in the context of a fire claim, see *Mueller Europe Ltd v Central Roofing (South Wales) Ltd* [2013] EWHC 237 (TCC), 147 Con. L.R. 32.

The "but for" test.

9–062 [Add a new sentence after: "Subject to these exceptions ... it is likely that there will be a requirement to satisfy the "but for" test.": page 353]

It may be, however, that in certain circumstances, the burden of proving that the loss in question would have occurred in any event, rests on the defendant.[275A]

[275A] In *West v Ian Finlay & Associates* [2013] EWHC 868 (TCC) where the defendant architect was found to have been negligent in failing to notice the poor quality of the M&E services installations in the renovation of the claimants' property, but argued that there was no loss because the work under this contract would never have been properly carried out and completed by the M&E contractor irrespective of how competently it had acted. It was held that given the nature of its breaches of duty, the onus was on the defendant to show that even if it had acted with reasonable care, the damage would probably still have occurred, applying the approach adopted by the

Court of Appeal in the context of a road traffic personal injury claim in *Phethean-Hubble v Coles* [2012] EWCA Civ 349, [2012] R.T.R. 31. Overturned on other grounds in *West v Ian Finlay and Associates* (2014) 153 Con. L.R. 1; [2014] EWCA Civ 316.

Concurrent independent causes of loss: claimant and defendant.

[Add to note 294: page 357] **9–070**

Cf. John Marrin QC, "Concurrent Delay Revisited", Society of Construction Law Paper 179, February 2013, pp.17–19.

CHAPTER TEN

LIQUIDATED DAMAGES AND PENALTIES

1. Generally   10–001
2. Defences to claim for liquidated damages   10–004
   - ☐ (a) *Agreed sum is a penalty*   10–004
   - (b) *Omission of date in contract*   10–011
   - (c) *Waiver*   10–012
   - (d) *Final certificate*   10–013
   - (e) *Employer causing delay*   10–014
   - (f) *Breach of condition precedent by employer*   10–018
   - (g) *Extension of time*   10–019
   - ■ (h) *Rescission or determination*   10–020

2. DEFENCES TO CLAIM FOR LIQUIDATED DAMAGES

(a) Agreed sum is a penalty

Lord Dunedin's propositions.

[Add to note 35: page 365]

10–005

In certain circumstances the relevant date for assessment may be the date on which the agreement was varied, such as by a reduction in the contract price: see *Unaoil Ltd v Leighton Offshore Pte Ltd* [2014] EWHC 2965 (Comm), [71].

[Add to note 39: page 366]

For a recent example where a liquidated damages clause was found to be a penalty because any delay to practical completion was incapable of having any adverse financial consequences, see *Spiers Earthworks v Landtec Projects* [2012] B.L.R. 223 (WASC). See also *Unaoil Ltd v Leighton Offshore Pte Ltd* [2014] EWHC 2965 (Comm), where a liquidated damages provision became a penalty on the reduction of the contract price, from which point it could no longer be said to be a genuine pre-estimate of loss.

CHAPTER TEN—LIQUIDATED DAMAGES AND PENALTIES

[Add new paragraph to text after note 41: page 367]

However, it must also be observed that the meaning of the word "unconscionable" in Lord Dunedin's test has developed beyond the point where it is purely synonymous with "extravagant". A string of modern authorities, commencing with Colman J.'s judgment in *Lordsvale Finance Plc v Bank of Zambia*, supports the proposition that a clause providing for extravagant liquidated damages may nonetheless be valid if it is, for some other reason, commercially justifiable.[41A] In *Makdessi v Cavendish Square Holdings BV*[41B] Christopher Clarke L.J. comprehensively reviewed this area of the law. At the time of writing, the decision is the subject of an appeal to the Supreme Court, but three propositions can be derived from that case:

1. Where there is a range of possible loss attributable to the breach upon which liquidated damages are payable, the following guideline assist in determining whether the clause in question is a genuine pre-estimate of loss:

    "i) A sum will be penal if it is extravagant in amount in comparison with the maximum conceivable loss from the breach;
    ii) A sum payable on the happening or non happening of a particular event is not to be presumed to be penal simply because the fact that the event does or does not occur is the result of several breaches of varying severity;
    iii) A sum payable in respect of different breaches of the same stipulation is not to be presumed to be penal because the effect of the breach may vary;
    iv) The same applies in respect of breaches of different stipulations if the damage likely to arise from those breaches is the same in kind;
    v) But a presumption may arise if the same sum is applicable to breaches of different stipulations which are different in kind;
    vi) There is no presumption that a clause is penal because the damages for which it provides may, in certain circumstances, be larger than the actual loss; and
    vii) Where there is a range of losses and the sum provided for is totally out of proportion to some of them the clause may be penal."[41C]

2. Christopher Clarke L.J. noted, what he referred to as the new approach (in contrast to the approach suggested by the adoption of the *Dunlop* propositions), where the court adopts a "broader test of whether the clause was extravagant and unconscionable with a predominant function of deterrence; and robustly declining to do so in circumstances where there was a commercial justification for the clause. That this is a reversion to the foundation of the doctrine is

apparent from the observations of Lord Halsbury in Clydebank Engineering when he asked of the relevant clause: 'whether it is, what I think gave the jurisdiction to the Courts in both countries to interfere at all in an agreement between the parties, unconscionable and extravagant, and one which no Court ought to allow to be enforced'."[41D]

3. Christopher Clarke L.J. summed up the effect of this development on the meaning of the word "unconscionable" in the following terms:

"Nowadays, when a term which provides for excessive payment on breach may be valid if it has a proper commercial justification, the term 'unconscionable' would, perhaps more appropriately be used for a clause which provides for extravagant payment without sufficient commercial justification. Such a clause is likely to be regarded as penal and deterrence its predominate function, on the basis that if it requires excessive payment, and lacks commercial justification for doing so, there is little room for any conclusion other than its function is to deter breach or, to put it positively, to secure performance."[41E]

[41A] See *Lordsvale Finance Plc v Bank of Zambia* [1996] Q.B. 752, 764, endorsed by the Court of Appeal in *Cine Bes Filmcilik Ve Yapim Click v United International Pictures* [2003] EWCA Civ 1669, per Mance L.J., at [15]; and in *Murray v Leisureplay Plc* [2005] EWCA Civ 963, per Arden L.J., at [46] and [54].
[41B] [2014] B.L.R. 246; [2013] EWCA Civ 1539.
[41C] At [71].
[41D] At [104].
[41E] At [125].

Sectional completion and partial possession.

[Add to note 51: page 368]:   10–006

For an example of a contract which permitted the employer to choose to levy liquidated damages at a reduced rate and thereby potentially save the contractual rate from being struck down as a penalty, see *Lidl UK GmbH v RG Carter Colchester Ltd* (2012) 146 Con. L.R. 133.

Set-off clauses

[Add to note 67: page 370]   10–009

Cf. a recent decision of the Australian High Court, *Andrews v Australia and New Zealand Banking Group Ltd* [2013] B.L.R. 111; [2012] HCA 30, where it was held that a clause requiring payment of money other than for a breach of contract could be characterised as a penalty. It is not thought this authority would be followed, on similar facts, in England and Wales.

Damages where the agreed sum is a penalty.

11–010 [Amend reference in note 75: page 371]

as cited at fn.114 to para.15–027 of H. McGregor, *McGregor on Damages*, 19th edn (London: Sweet & Maxwell, 2014).

[Amend reference in note 77: page 371]

McGregor on Damages, 19th edn, (London: Sweet & Maxwell, 2014) comes to the opposite conclusion at para.15–027.

(h) Rescission or determination

10–020 [Add to note 104 after the reference to *Suisse Atlantique v N.V. Rotterdamsche, etc.* [1967] A.C. 361, HL: page 376]

LW Infrastructure v Lim Chin San [2012] B.L.R. 13 (High Court of Singapore),

CHAPTER ELEVEN

DEFAULT OF THE PARTIES—VARIOUS MATTERS

1.	Forfeiture clauses	11–001
	(a) *The nature of forfeiture clauses*	11–002
	(b) *The mode of forfeiture*	11–003
	■ (c) *Employer's default*	11–006
	■ (d) *Effect of forfeiture*	11–007
2.	Materials and plant	11–013
	☐ (a) *Ownership of materials and plant*	11–013
	■ (b) *Vesting clauses*	11–015
3.	Lien	11–018
4.	Defects and maintenance clauses	11–019
	(a) *Meaning of terms*	11–020
	(b) *Investigation for defects*	11–021
	(c) *Notice is required*	11–022
	(d) *Alternative claim in damages*	11–023
	(e) *Liability after expiry of period*	11–024
5.	Guarantees and bonds	11–025
	☐ (a) *Guarantees*	11–025
	☐ (b) *Bonds*	11–034
☐ 6.	Liability to third parties	11–038
	■ (a) *Who is liable?*	11–039
	☐ (b) *Nuisance from building operations*	11–041
7.	Contractor's duty of care towards employer	11–047
8.	Claim for breach of confidence	11–050

CHAPTER ELEVEN—DEFAULT OF THE PARTIES—VARIOUS MATTERS

1. FORFEITURE CLAUSES

(c) Employer's default

11–006 [Add to note 34: page 381]

Contrast *Inframatrix Investments Ltd. v Dean Construction Ltd.* [2012] EWCA Civ 64, [2012] 2 All E.R. (Comm) 337 in which the principle in *Alghussein* was held not to be applicable to a contractual time-bar provision.

(d) Effect of forfeiture

Relief against forfeiture.

11–011 [Add to note 50: page 383]

and *Cavendish Square Holdings BV v Makdessi* [2012] EWHC 3582 (Comm), [41], this point not affected by the appeal at [2013] EWCA Civ 1539.

[At note 51, after the reference to *BICC v Burndy Corp* [1985] Ch. 232, CA, add: page 383]

Cukorova Finance International Ltd v Alfa Telecom Turkey Ltd [2013] UKPC 2;

2. MATERIALS AND PLANT

(a) Ownership of materials and plant

Retention of title clauses.

11–014 [Add at end of note 71: page 386]

For a similar scenario, in which a retention of title clause had the effect that the purchaser's resale to its subsidiary was effected as the seller's fiduciary agent, see *F.G. Wilson (Engineering) Ltd v John Holt & Co (Liverpool) Ltd* [2014] 1 W.L.R. 2365; [2013] EWCA Civ 1232.

(b) Vesting clauses

11–015 [Add to note 79: page 386]

See also, *Alstom Power Ltd v Somi Impianti Srl* [2012] EWHC 2644 (TCC).

5. GUARANTEES AND BONDS

(a) Guarantees

[Amend reference in note 119: page 392] 11–025

see *Chitty on Contracts*, edited by H. Beale, 31st edn (London: Sweet & Maxwell, 2012), Vol.2, Ch.44;

Liability of the surety.

[After the words "had no knowledge" delete the full stop and (following 11–026
note 123) insert: page 392]

or to monies paid under a separate agreement to pay sums in respect of claims for delay, which amounted to a loan.[123a]

[123A] *Hackney Empire Ltd v Aviva Insurance UK Ltd* [2012] EWCA Civ 1716, 146 Con. L.R. 1.

Non-disclosure.

[Amend reference in note 139: page 394] 11–031

see also, *Chitty on Contracts*, edited by H. Beale, 31st edn (London: Sweet & Maxwell, 2012), Vol.2, para.44–035.

Material alteration in contract.

[At note 158 the case of *Hackney Empire Ltd v Aviva Insurance UK Ltd* is 11–033
now reported at: page 396]

[2012] EWCA Civ 1716, 146 Con. L.R. 1.

(b) Bonds

Conditional bonds.

[At note 176, following the reference to *Carey Value Added SL v Grupo* 11–035
Urvasco (2010) 132 Con. L.R. 15, add: page 399]

and *Wuhan Guoyo Logistics Group v Emporiki Bank of Greece SA* [2012] EWCA Civ 1629, [2013] 1 All E.R. (Comm) 1191.

"On-demand bonds".

[At note 179, following the reference to *Carey Value Added SL v Grupo* 11–036
Urvasco (2010) 132 Con. L.R. 15, add: page 399]

Wuhan Guoyo Logistics Group v Emporiki Bank of Greece SA [2012] EWCA Civ 1629, [2013] 1 All E.R. (Comm) 1191, in each case the court finding that, on the true construction of the document, it was in the nature of a guarantee.

[At note 179, following the reference to *Meritz Fire & Marine Insurance Co Ltd v Jan de Nul NV* [2010] EWHC 3362 (Comm), (2010) 134 Con. L.R. 252, add]:

The decision was upheld on appeal [2011] EWCA 827, [2012] 1 All E.R. (Comm) 182. See also, *WS Tankship II BV v The Kwanju Bank Ltd*. [2011] EWHC 3103.

[Add to note 182: page 400]

See also *Doosan Babcock Ltd v Comercializadora de Equipos y Materiales Mabe Lda* [2014] B.L.R. 33; [2014] EWHC 3201 (TCC), where an injunction was granted to restrain a call on the bond where its continuing validity was alleged to be the result of the beneficiary's breach of contract.

6. LIABILITY TO THIRD PARTIES

11–038 [Add to note 192: page 401]

Barr v Biffa Waste Services Ltd. [2012] EWCA Civ 312; *Northumbrian Water Ltd v Sir Robert McAlpine Ltd* (2014) 154 Con. L.R. 26, [2014] EWCA Civ 685.

(a) Who is liable?

Employer.

11–040 [Add to note 211: page 404]

See also, *Tinseltime v Roberts* [2011] EWHC 1199; [2011] B.L.R. 515.

(b) Nuisance from building operations

Trees.

11–044 [Add to note 236: page 407]

Berent v Family Mosaic Housing [2012] EWCA Civ 961.

Pipelines.

11–046 [Substitute note 243: page 408]

Bocardo SA v Star Energy UK Onshore Ltd [2011] 1 A.C. 30; [2010] UKSC 35.

CHAPTER TWELVE

VARIOUS EQUITABLE DOCTRINES AND REMEDIES

☐ 1.	Estoppel	12–001
☐ 2.	Waiver	12–005
☐ 3.	Variation and rescission	12–009
☐ 4.	Rectification	12–011
☐ 5.	Specific performance	12–017
☐ 6.	Injunction	12–020

1. ESTOPPEL

Estoppel is an equitable doctrine.

[Amend reference in note 2: page 411] **12–001**

For a full account, see, e.g. *Halsbury's Laws of England*, 5th edn (London: LexisNexis, 2014), Vol.47; [*Estoppel by Representation* . . .].

Estoppel by representation.

[Substitute second sentence of paragraph, and add new note 5A: page 412] **12–002**

The representation may be a representation of fact or of an existing state of mind or belief but must not be a future promise.[5] It also now appears that the representation may be a statement of law.[5A]

[5A] *Re Gleeds Retirement Benefits Scheme* [2014] Pens. L.R. 265; [2014] EWHC 1178 (Ch), applying the decision of the House of Lords in *Kleinwort Benson Ltd v Lincoln CC*[1999] 2 A.C. 349.

[Add to note 7: page 412]

PCE Investors Ltd v Cancer Research UK [2012] 2 P.&C.R. 5 (Ch.) in which it was confirmed (obiter) that there was no general proposition of law that where one party perceived the other side was making a mistake they had a duty to correct it.

CHAPTER TWELVE—VARIOUS EQUITABLE DOCTRINES AND REMEDIES

[Add the following text after the sentence: "It was held ... the defendant's limitation defence": page 412]

Where a party to a contract to repair had represented that there was no need to amend the contract to reflect the party's prior agreement on the measure of rates, it was reasonably arguable that estoppel and misrepresentation might apply, notwithstanding the existence of an entire agreement clause.[10A]

[10A] *Mears Ltd v Shoreline Housing Partnership Ltd* [2013] EWCA Civ 639.

Estoppel by convention.

12–003 [Amend reference in note 11: page 413]

See also, *Chitty on Contracts*, edited by H. Beale, 31st edn (London: Sweet & Maxwell, 2012), Vol.1, para.3–107 and following.

[Add the following text after the sentence "Once a common assumption ... will not apply to future dealings.": pages 412–413]

However, no estoppel arose where the evidence was more consistent with ordinary business co-operation than with reliance on and recognition of a shared common assumption of a legal obligation.[11A] Nor was there an estoppel where the parties to a contract had operated it as though a similarly named third party were the contractor; in doing so, they did not adopt a convention that the third party replaced its namesake as a party to the contract.[11B]

[11A] *Jet2.com Ltd v Blackpool Airport Ltd* [2012] 142 Con. L.R. 1 (CA).
[11B] *Liberty Mercian Ltd v Cuddy Civil Engineering Ltd* (2013) 150 Con. L.R. 124: [2013] EWHC 2688, [152]–[165].

Promissory estoppel.

12–004 [Add to the sentence (ahead of note 17) at the end of the paragraph, "The promise need not be supported by consideration": page 414]

, but it must be shown that the promise or assurance had a sufficiently material influence on the other party's conduct to make it inequitable for the promisor to depart from it.[17A]

[17A] *Crossco No.4 Unlimited v Jolan Ltd* [2011] N.P.C 38 (Ch.).

2. WAIVER

12–005 [Amend reference in note 20: page 414]

see also cases cited in *Chitty on Contracts*, edited by H. Beale, 31st edn (London: Sweet & Maxwell, 2012), Vol.1, para.22–042.

3. VARIATION AND RESCISSION

[Amend reference in note 37: page 416]   12–009

see generally, *Chitty on Contracts*, edited by H. Beale, 31st edn (London: Sweet & Maxwell, 2012), Vol.1, para.22–032 and following.

4. RECTIFICATION

Common mistake.

[Add to text after note 41: page 417]   12–012

Where the instrument to be rectified simply fails to deal with the matter in question, the court cannot rectify a non-existent term. In such cases, the better approach is to deal with the matter as an implied term.[41A]

[41A] *OMV Supply and Trading AG v Kazmunaygaz Trading AG* [2014] EWHC 1372 (Comm), [70]–[82]. As to implied terms generally, see above para.3–039 and following.

[Add to note 47: page 418]:

and in *Scottish Widows Fund and Life Assurance Society v BGC International* [2012] 142 Con. L.R. 27 (CA). See also *Liberty Mercian Ltd v Cuddy Civil Engineering Ltd* (2013) 150 Con. L.R. 124, [2013] EWHC 2688.

[Add the following text after the sentence "Thus rectification has been granted ... understandings reached without agreement in words.[51]": page 418]

It must be clear from the rest of the agreement interpreted against the admissible background what the parties intended to agree and the mistake must be one of language or syntax.[51A]

[51A] *Scottish Widows Fund and Life Assurance Society v BGC International* [2012] 142 Con. L.R. 27 (CA) where the court held that it was not enough that the parties had mistakenly failed to provide for a particular circumstance, because were the court to correct such an error it would mean that it was re-writing the parties' contract for them which was impermissible. The court also emphasised that the burden of proving that the requirements for rectification had been fulfilled fell on the party seeking rectification and stated that the burden was more difficult to discharge where the instrument was detailed and had been drafted with the benefit of expert legal advice.

Unilateral mistake.

12–013 [Substitute note 54: page 419]

In *George Wimpey UK Ltd v VIC Construction Ltd* [2005] B.L.R. 135, CA, it was held that the knowledge of the other party must be either: (1) actual; (2) the wilful shutting of one's eyes to the obvious; or (3) the wilful and reckless failure to make such inquiries as an honest and reasonable person would make.

[Add the following text as a new paragraph after "In these circumstances ... was not a mutual mistake.⁵⁶": page 419]

In determining the intentions of the parties as to the meaning of the contract, it is necessary to look at the intentions of the person with authority to bind that party to the contract rather than a mere negotiator.[56A]

[56A] *Hawksford Trustees Jersey Ltd v Stella Global UK Ltd* [2012] 2 All E.R. (Comm) 748 (CA).

[Add a new paragraph after para.12–015: page 421]

Expert determination.

12–015A Whether a clause referring a dispute to expert determination covers claims for rectification claims is likely to depend on the wording of the clause. However, there is recent authority suggesting that the remedy may by its nature be outside of the scope of expert determination.[75A]

[75A] *Persimmon Homes Ltd v Woodford Land Ltd* [2012] B.L.R. 73 (Ch.), [21] but note that the point appears to have been conceded rather than the subject of full argument.

5. SPECIFIC PERFORMANCE

12–017 [Add the following to note 81: page 422]

XY v Facebook Ireland Ltd [2012] NIQB 96.

Impossibility.

12–017A If the obligation in question is incapable of being performed this does not necessarily prevent the court making an appropriate order. Thus in *Liberty Mercian Ltd v Cuddy*,[84A] following termination of the contract, the contracting party under an amended NEC3 Form of Contract, was ordered to use its best endeavours to obtain the performance bond and collateral warranties as specified by the contract. In the event, the contracting party was unable to obtain such a bond and warranties. As regards the performance bond, the court held it could order substituted performance pursuant to its equitable jurisdiction utilising the procedural machinery of payment into court granted by s.19 of the Senior Courts Act 1981 to stand as equivalent to the performance bond on the same terms and conditions as would have applied had a performance bond been provided as required by the contract.[84B] The court also considered whether substituted performance could be provided by way of a bond, the terms of which were not as advantageous as the terms specified by the contract. The court did not think provision of this bond was appropriate substituted performance since it "would not be one which a prudent party would enter into." As regards the collateral warranties, the company which was obliged to provide them had gone into liquidation. However there was some evidence that the company might have relevant insurance cover. Notwithstanding this uncertainty, the court ordered specific performance of the obligation to provide the collateral warranties.

[84A] [2013] EWHC 4110 (TCC).
[84B] [2014] EWHC 3584 (TCC).

6. INJUNCTION

[Add the following to note 91: page 424] 12–020

cf. *Qimonda Malaysia SDN BHD v Sediabena SDN BHD* [2012] B.L.R. 65 CA (Mal)

[Insert into third sentence of first paragraph: page 424]

But there are a number of common situations arising under building contracts... to restrain a call on a bond,[91A] to prevent the use of plans produced by the applicant,[91B] or in order to give effect to [contractual provisions for ...

[91A] *Simon Carves Ltd v Ensus UK Ltd* (2011) 135 Con. L.R. 96, [2011] EWHC 657 (TCC); *Doosan Babcock Ltd v Comercializadora de Equipos y Materiales Mabe Lda* [2014] 1 Lloyd's Rep. 464, [2013] EWHC 3010 (TCC); cf. *Seele Middle East FZE v Raiffeisenlandesbank Oberosterreich Aktiengesellschaft Bank* [2014] EWHC 343 (TCC).
[91B] *Seele Middle East FZE v Drake and Scull International SA Co* [2014] EWHC 435 (TCC).

[Add new sentence after "A claim to an interim injunction cannot exist in isolation.[98]": page 424]

The court is unlikely to grant an injunction to restrain interference with a party's rights unless the threat is tangible and imminent.[98A]

98A *CIP Property (AIPT) Ltd v Transport for London* [2012] B.L.R. 202 (Ch.) where an injunction to prevent infringement of a right to light was refused because there was no immediate threat.

Injunction to enforce disputed forfeiture.

12–021 [Add to note 106: page 425]

Hoad & Taylor v Delves [2012] EWHC 1426 (QB).

[Add to note 109: page 426]

and *Financial Services Authority v Sinaloa Gold Plc* [2013] 2 W.L.R. 678 (SC).

Chapter Thirteen

ASSIGNMENTS, SUBSTITUTED CONTRACTS AND SUB-CONTRACTS

☐ 1.	Assignments	13–001
■ (a)	*Assignment by contractor of burden*	13–002
☐ (b)	*Assignment by contractor of benefit*	13–005
(c)	*Assignment by employer of burden*	13–017
☐ (d)	*Assignment by employer of benefit*	13–018
2.	Substituted contracts	13–024
3.	Sub-contractors	13–025
(a)	*Liability for sub-contractors*	13–025
(b)	*Relationship between sub-contractors and employer*	13–027
(c)	*Prime cost and provisional sums*	13–034
4.	Nominated sub-contractors and nominated suppliers	13–035
(a)	*Delay in nomination*	13–036
(b)	*Problems inherent in nomination*	13–037
(c)	*Express terms of main contract*	13–038
(d)	*Implied terms of main contract*	13–041
(e)	*Repudiation by nominated sub-contractor*	13–051
(f)	*Protection of employer*	13–052
☐ 5.	Relationship between sub-contractor and main contractor	13–054

1. ASSIGNMENTS

Assignments.

[Amend reference in note 1: page 428] **13–001**

see *Chitty on Contracts*, edited by H. Beale, 31st edn (London: Sweet & Maxwell, 2012). Vol.1, Ch.19.

CHAPTER THIRTEEN—ASSIGNMENTS, SUBSTITUTED CONTRACTS AND SUB-CONTRACTS

Vicarious performance.

13–004 [Add following text to note 8 after: "*Nokes v Doncaster Amalgamated Collieries Ltd* [1940] A.C. 1014, HL;": page 429]

Cf. *North v Brown* [2012] EWCA Civ 223, CA confirming the distinction between assignment and vicarious performance drawn in *Nokes* (at [12]).

[Add the following text at the end of the final paragraph of 13–004, after the sentence: "A contract may be construed ... a £100 company": page 431]

Indeed the obligation to pay money under a contract does not generally depend on the personality of the paying party and so will usually be capable of vicarious performance even if assignment is not permitted.[21A]

[21A] *North v Brown* [2012] EWCA 223, CA.

(b) Assignment by contractor of benefit

13–005 [Add to note 30 following the reference: *Three Rivers DC v Bank of England* [1996] Q.B.292, CA: page 432]

followed in *Bexhill UK Ltd v Razzaq* [2012] EWCA Civ.1376 and *National Westminster Bank Plc v Kapoor* [2011] EWCA Civ. 1083.

[Add to note 31: page 432]

National Westminster Bank Plc v Kapoor [2011] EWCA Civ 1083, [30].

Statutory assignment.

13–006 [Add to note 42 following the reference *Rayack Construction Ltd v Lampeter Meat* (1979) 12 B.L.R. 30: page 433]

cf. *Qimonda Malaysia SDN BHD v Sediabena SDN BHD* [2012] B.L.R. 65 CA (Mal) where part of the reasoning was disapproved;

Equitable assignment.

13–007 [At note 52 add after the reference *Rayack Construction Ltd v Lampeter Meat* (1979) 12 B.L.R. 30: page 434]

cf. *Qimonda Malaysia SDN BHD v Sediabena SDN BHD* [2012] B.L.R. 65 CA (Mal) where part of the reasoning was disapproved;

Fraud.

13–013 [Substitute last sentence of note 81: page 437]

See commentary on this case, *Chitty on Contracts*, edited by H. Beale, 31st edn (London: Sweet & Maxwell, 2012), Vol.1, para.19–071.

Assignments

(d) Assignment by employer of benefit

Effect of invalid assignment.

[Add sentence at end of paragraph: page 438]   13–019

An invalid assignment will not give rise to a trust in favour of the assignee, with a possible exception where the language of the attempted assignment can be construed as a declaration of trust.[93A]

[93A] *Co-operative Group Ltd v Birse Developments Ltd (In Liquidation)* (2014) 153 Con. L.R. 103; [2014] EWHC 530 (TCC), [61]–[92].

Maintenance, champerty and assignment of causes of action.

[Add to note 94: page 439]:   13–020

Simpson v Norfolk and Norwich University Hospital NHS Trust [2011] EWCA Civ 1149; *Skywell (UK) Ltd v Revenue and Customs Commissioners* [2012] UKFTT 61 (TC).

[Add to note 95: page 439]

Simpson v Norfolk & Norwich University Hospital NHS Trust [2011] EWCA Civ 1149 [23–24] and [25] where the court rejected the submission that a refusal to recognise an assignment of a bare right to litigate was an unlawful interference of the claimant's property contrary to art.1 of the European Convention on Human Rights.

[At note 97 add a new reference after *Brownton v Edward Moore Inbucon* [1985] 3 All E.R. 499, CA: page 439]

Golden Eye (International) Ltd v Telefonica UK Ltd [2012] EWCA Civ 1740.

Assignment of warranties.

[Add to note 102: page 440]   13–021

Pegasus v Ernst & Young [2012] EWHC 738 (Ch).

Contracts (Rights of Third Parties) Act 1999.

[Amend reference in note 119: page 442]   13–022

See *Chitty on Contracts*, edited by H. Beale, 31st edn (London: Sweet & Maxwell, 2012), Vol.1. paras 18–088 to 18–120.

[Add new paragraph at end of the final paragraph of 13–022 after the sentence, "Further, the promise ... the third party had been a party to the contract": page 443]

Whilst the statutory language appears to envisage that a third party may be bound by an arbitration clause, it would need very clear language to bring it about.[124A] It does not appear that third parties can take the benefit of an adjudication clause, regardless of the language used.[124B]

[124A] *Fortress Valley Recovery Fund I LLC v Blue Skye Special Opportunities Fund LP* [2013] EWCA 367.
[124B] *Hurley Palmer Flatt Ltd v Barclays Bank plc* [2014] EWHC 3042 (TCC), [42].

5. RELATIONSHIP BETWEEN SUB-CONTRACTOR AND MAIN CONTRACTOR

13–054 [Add to note 218: page 458]

; and *Quashie v Stringfellow Restaurants Ltd* [2012] EWCA Civ 1735.

Sub-contractor's access to site.

13–057 [Add a new footnote at the end of sentence "Otherwise it is likely ... as will enable it properly to perform its sub-contract obligations.": page 461]

[236A] There is Australian authority to the effect that, in the absence of express terms, there may be an implied obligation for the contractor and sub-contractor to act in good faith if their respective contractual obligations demand a high degree of co-operation and reliance upon each other's good faith: *Alstom Ltd v Yokogawa Australia Pty Ltd* (No.7) [2012] SASC 49.

Set-off.

13–058 [Add to note 243: page 462]

For the position on the right of an unsuccessful party to set-off claims against a decision of an adjudicator see: *Squibb Group Ltd v Vertase FLI Ltd* [2012] B.L.R. 408, [2012] EWHC 1958 (TCC); and *Thameside Construction Co Ltd v Stevens* (2013) 149 Con. L.R. 195, [2013] EWHC 2071.

Damages.

13–060 [Add to note 253: page 463]

John Grimes Partnership Ltd v Gubbins [2013] EWCA Civ 37.

[Add at note 254 after the reference *"Transfield Shipping Inc v Mercator Shipping Inc* [2008] W.L.R. 345, HL": page 463]

cf. *John Grimes Partnership Ltd v Gubbins* [2013] EWCA Civ 37, [20];

Chapter Fourteen

ARCHITECTS, ENGINEERS AND SURVEYORS

1.	Introduction	14–001
2.	Meaning and use of the term "architect"	14–002
☐ 3.	Registration	14–003
4.	The position of the architect	14–010
	(a) *The contract with the employer*	14–010
	☐ (b) *Architects' authority as agent*	14–013
	☐ (c) *Excess of authority by architect*	14–020
	(d) *Architects' personal liability on contracts*	14–024
	(e) *Fraudulent misrepresentation*	14–025
	☐ (f) *Misconduct as agent*	14–026
	■ (g) *Architects' duties to the employer*	14–028
	☐ (h) *Architects' duties in detail*	14–035
	■ (i) *Breach of architects' duties to employer*	14–055
	(j) *Duration of architects' duties*	14–062
	(k) *Remuneration*	14–066
	(l) *The architects' lien*	14–073
	(m) *Property in plans and other documents*	14–075
	(n) *Copyright in plans and design*	14–078
	(o) *Architects' duties to contractors and others*	14–083
5.	Engineers and others	14–088
6.	Quantity surveyors	14–093
	(a) *Surveyors' duties generally*	14–093
	(b) *Architects' authority to engage quantity surveyor*	14–094
	(c) *Surveyors' duties to employer*	14–098
	(d) *Remuneration*	14–101
	(e) *Duties to contractor and others*	14–103

CHAPTER FOURTEEN—ARCHITECTS, ENGINEERS AND SURVEYORS

3. REGISTRATION

Retention of name on Register.

14–007 [Add to note 27: page 469]

Where re-admission is refused on grounds other than competence, the appropriate route of appeal is by judicial review: *Dowland v Architects Registration Board* [2013] B.P.I.R. 566; [2013] EWHC 893 (Admin).

Professional conduct.

14–008 [Delete second sentence of note 38: "For procedure see CPR Pt 52 and s.III of the Practice Direction to Pt 52—Provisions about Specific Appeals" and substitute: page 470]

For procedure see CPR Pt 52, PD52A—Appeals and PD52D—Statutory appeals and appeals subject to special provision, Section IV—Specific Appeals, in particular para.19.1 of PD52D.

4. THE POSITION OF THE ARCHITECT

(b) Architect's position as agent

14–014 [Amend reference in note 56: page 472]

Chitty on Contracts, edited by H.G. Beale, 31st edn (London: Sweet & Maxwell, 2012), Vol.2, Ch.31.

(c) Excess of authority by architect

Position of the employer.

14–020 [Amend reference in note 75: page 474]

Chitty on Contracts, edited by H.G. Beale, 31st edn, (London: Sweet & Maxwell, 2012), Vol.2, para.31–057.

[Amend reference in note 81: page 475]

see *Chitty on Contracts*, edited by H.G. Beale, 31st edn (London: Sweet & Maxwell, 2012), Vol.2, para.31–096;

14–021 [Amend reference in note 82: page 475]

See further, *Chitty on Contracts*, edited by H.G. Beale, 31st edn (London: Sweet & Maxwell, 2012), Vol.2, para.31–026.

(f) Misconduct as agent

Bribes and secret commissions.

[Add to end of note 109: page 479] **14–026**

Prevention of Corruption Act 1916 s.2 was repealed by the Bribery Act 2010.

[Add at the end of note 114: page 479]

In *FHR European Ventures LLP v Mankarious* [2013] EWCA Civ 17, [2013] 3 W.L.R. 466 where a consultancy advised and negotiated on behalf of the purchasers of a hotel but failed to disclose a commission received from the vendors of the hotel, it was held that the purchaser's remedy included the ability to trace into the commission paid which the consultancy held on a constructive trust for the investor group.

(g) Architects' duties to the employer

Delegation of duties.

[Insert a new paragraph at the end of 14–033 following: "In view of successful work done elsewhere the architect's decision to use Pyrok was reasonable and no witness was called to suggest that it was not reasonable.[161]": page 486] **14–033**

Construction professionals did not by the mere act of obtaining advice or a design from another party thereby divest themselves of their duty in respect of that advice or design. They could discharge their duty to take reasonable care by relying on the advice or design of a specialist provided that they acted reasonably in doing so. In determining whether construction professionals acted reasonably in seeking the assistance of specialists to discharge their duty to a client, the court had to consider all of the circumstances which including whether:

(a) the assistance was taken from an appropriate specialist;

(b) it was reasonable to seek assistance from other professionals, research or other associations or other sources;

(c) there was information which should have led the professional to give a warning;

(d) the client might, and to what extent, have a remedy in respect of the advice from the other specialist; and

(e) the construction professional should have advised the client to seek advice elsewhere or should themselves have taken professional advice under a separate retainer.[161A]

[161A] *Cooperative Group v John Allen Associates* (2012) 28 Const. L.J. 27.

CHAPTER FOURTEEN—ARCHITECTS, ENGINEERS AND SURVEYORS

(h) Architects' duties in detail

Law and practice.

14–037 [Add at end of note 184: page 490]

Where a multi-discipline consultancy of engineers, architects and planners advised that a new pumping station was a permitted development not requiring planning permission and did not recommend the seeking of the Certificate of Lawful Development, they were held not to be negligent: *Middle Level Commissioners v Atkins Ltd* [2012] EWHC 2884 (TCC).

Estimates.

14–039 [Add to note 191: page 490]

See also *Pickard Finlason Partnership Ltd v Lock* [2014] EWHC 25 (TCC), where a design consultancy's failure to provide cost information at the feasibility stage, in the hope that costs savings could be identified to increase the scheme's chances of being approved, was held to be a breach of contract.

Advising on the contract.

14–044 [Add a new sentence at the end of the final paragraph of 14–044 after: "Failure to take such steps as are referred to in the preceding two paragraphs may furnish some prima facie evidence of negligence, but each case would depend on its own facts.": page 495]

A failure to take the steps necessary to finalise a contract but instead to rely on the issue of repeated letters of intent may also be a breach of duty.[230A]

[230A] *The Trustees of Ampleforth Abbey Trust v Turner & Townsend Project Management Limited* [2012] EWHC 2137 (TCC) in relation to project managers who acted as "the representative of the employer for the purpose of co-ordinating the different aspects of a construction or engineering project" or as "co-ordinator and guardian of the client's interests".

(i) Breach of architects' duties to employer

14–055 [Insert a new sentence at the end of paragraph 14–055 following: "But where a design was found to be defective...had the new design been incorporated originally.": page 503]

In some cases the failure to act may give rise to damages based on the loss of a chance.[280A]

[280A] In *The Trustees of Ampleforth Abbey Trust v Turner & Townsend Project Management Limited* [2012] EWHC 2137 (TCC) the project managers were in breach of duty in failing to take the steps necessary to finalise a contract with the contractor. It was held that there was a real and substantial chance that any such contract would have contained a provision for liquidated damages. The court made an assessment of the loss suffered by the employer based on evidence of the damages provision which might have been incorporated into a final contract and a chance that the contractor would have signed a contract incorporating such a term assessed, at two-thirds.

When does the cause of action arise?

[Add at the end of note 298: page 506] **14–060**

A continuing duty for a firm of solicitors to review all advice and drafting it had previously carried out for a construction company was hopelessly wide: *Shepherd Construction Ltd v Pinsent Masons LLP* [2012] EWHC 43 (TCC).

Chapter Fifteen

PUBLIC PROCUREMENT

☐ 1. European community legislation 15–001
■ 2. Common law remedies relating to public procurement 15–044
■ 3. Local government procurement legislation 15–046

1. EUROPEAN COMMUNITY LEGISLATION

Introduction.

[Amend references in note 2: page 526] **15–001**

Directive 18/2004/18 art.2, Directive 2014/24/EU art.18 and the Public Contracts Regulations 2006 (SI 2006/5) reg.4(3)
Directive 17/2004 art.10, Directive 2014/25/EU art.36 and the Utilities Contracts Regulations 2006 (SI 2006/6) reg.4(3).
The 2014 Directives came into force on the 20th day following their publication on March 28, 2014, and the 2004 Directives will be repealed with effect from April 18, 2016. Pending the coming into force of the domestic legislation implementing the 2014 Directives, the 2004 Directives will continue to be effective.

The Directives and Regulations.

[Substitute first paragraph: page 527] **15–002**

Consolidated and updated Directives on public procurement were adopted in February 2014. These were Directive 2014/24 (relating to public contracts), Directive 2004/25 (relating to utilities), and Directive 2004/23 (relating to concessions contracts).[3] The new Directives are intended to modernise procurement and to make the public procurement process faster, less costly and more effective. It is expected that Directive 2014/24 will be transposed into UK law in 2015.[3A] Until that transposition occurs, the Public Contracts Regulations 2006[4] ("the Public Contracts Regulations") and the Utilities Contracts Regulations 2006[5] ("the Utilities Contracts Regulations") remain in force and apply to contract award procedures commenced on or after January 31, 2006.[6]

³ Directive 2004/18, co-ordinating the procedures for the award of public works contracts, public supply contracts and public service contract and Directive 2004/17, co-ordinating the procurement procedures of entities operating in the water, energy, transport and postal services sectors.

³ᴬ The draft transposing regulations, published during consultation, can be found at: *https://www.gov.uk/government/uploads/system/uploads/attachment_ data/file/356494/Draft_Public_Contracts_Regulations_2015.pdf* [Accessed October 9, 2014].

⁴ (SI 2006/5) implementing Directive 2004/18.

⁵ (SI 2006/6) implementing Directive 2004/17.

⁶ See the Public Contracts Regulations 2006 (SI 2006/5) regs 1(1) and 49(1) and the Utilities Contracts Regulations 2006 (SI 2006/5) regs 1(1) and 48(1). The Regulations have been amended by the Public Contracts and Utilities Contracts (Amendment) Regulations 2007 (SI 2007/3542); the Public Contracts (Amendment) Regulations 2009 (SI 2009/2992); the Utilities Contracts (Amendment) Regulations 2009 (SI 2009/3100); and the Public Procurement (Miscellaneous Amendments) Regulations 2011 (SI 2011/2053).

[Add to text at end of note 15: page 528]

In *Insinööritoimisto InsTiimi Oy* (C-615/10), unreported, June 7, 2012, it was held that the exemption in art.346 applied to the procurement of war material which, although intended for military purposes, could also be used for civilian applications only if that material, by virtue of its intrinsic characteristics, might be regarded as having been specially designed and developed or substantially modified for military purposes.

[Replace note 18: page 528]

¹⁸ *https://www.gov.uk/government/publications/the-european-union-defence-and-security-public-contracts-regulations-dspcr-2011* [Accessed September 17, 2013].

Public works contracts.

15–005 [Add to end of note 24: page 530]

As also amended by the Planning Act 2008 and the Growth and Infrastructure Act 2013.

[Add new note 26A at end of sentence below: page 531]

In addition the Contractor must assume a direct or indirect obligation to carry out the works, which is legally enforceable.²⁶ᴬ

²⁶ᴬ In *R. (On the Application of Midlands Co-Operative Society Ltd) v Birmingham CC and Tesco Stores Ltd* [2012] EWHC 620 (Admin), it was

held that the sale of land to a supermarket was not a public works contract where the supermarket was not subject to a legally binding obligation to develop the land.

[Delete the full stop and add at end of note 22: page 529]

; and *Impresa Pizzarotti & C. SpA v Comune di Bari* (C-213/13).

Mixed contracts.

[Amend reference in note 36: page 532]   **15–008**

Commission v Germany (C-536/07) [2009].

[Amend reference in note 37: page 532]

Commission v Germany (C-536/07) [2009].

Contracting authorities.

[Add at the end of note 39: page 532]   **15–009**

In *Alstom Transport v Eurostar International Limited* [2012] EWHC 28 (Ch) it was held that Eurostar was not a contracting authority as it had a commercial character, given the introduction of competition into its market. Also, as it had received state aid, it was expected to operate going forward on a commercial basis.

[Add to note 40: page 532]

A public body financed mainly by contributions from its members, such as a national medical association, will not satisfy this criterion: *IVD GmbH & Co KG v Arztekammer Westfalen-Lippe* (C-526/11) [2014] 1 C.M.L.R. 31.

[Amend second sentence of note 41: page 532]

See also, *Commission v France* (C-237/99) [2001] E.C.R. I-939 and *IVD GmbH & Co. KG v Ørztekammer Westfalen-Lippe* (C-526/11) unreported, September 12, 2013.

Thresholds.

[Amend second sentence of note 44: page 533]   **15–010**

The current limits may be checked on the OJEC website: *http://www.ojec.com/Thresholds.aspx* [Accessed October 10, 2014].

CHAPTER FIFTEEN—PUBLIC PROCUREMENT

[Add at the end of note 48: page 533]

In *Germany v Commission* (T-258/06) [2010] the General Court dismissed a challenge to the Commission's Interpretative Communication, finding that it reflected the general Treaty principles as interpreted by ECJ case law.

Excluded contracts.

15–011 [Add at the end of note 49: page 534]

See *JBW Group Ltd v Ministry of Justice* [2012] EWCA Civ 8; *Photo-Me International PLC v Network Rail Infrastructure Limited* [2011] EWHC 3168 (QB).

[Add to note 56: page 535]

See also *Centro Hospitalar de Setúbal and SUCH v Eurest (Portugal)* (C-574/12).

[Add to note 59: page 535]

The same is true of a company limited by guarantee, of which a contracting authority is the sole member: *Tachie v Welwyn Hatfield BC* [2014] P.T.S.R. 662, [2013] EWHC 3972 (QB).

[Insert in fourth paragraph of para.15–011 following: "It is not necessary for the control to be exercised ... a number of authorities.⁶¹": page 535]

However, for joint control, each authority must not only hold capital in the entity, but also play a role in managing it.⁶¹ᴬ

⁶¹ᴬ *Econord SpA v Commune di Cagno* (C-182/11 & C-183/11) [2013] 2 C.M.L.R. 7.

[Insert new paragraph between fourth and fifth paragraphs: page 535]

As the control requirement would suggest, the *Teckal* exemption will normally arise in the context of vertical internal transactions, i.e. between a contracting authority and an entity it controls. One question that has yet to be decided is whether the exemption would apply to horizontal internal transactions, i.e. between two entities each controlled by same contracting authority. In such circumstances the *Teckal* exemption as formulated could not apply between the two parties to the contract, as neither entity would be in a position to control the other. However, it has been suggested that, so long as each entity is subject to such a degree of control by the contracting authority as would satisfy the control requirement for exemption in a vertical internal transaction, a horizontal internal transaction between those entities will similarly be exempt.⁶²ᴬ

82

⁶²ᴬ *Datenlotsen Informationssysteme GmbH v Technische Universität Hamburg-Harburg* (C-15/13) (Opinion of Advocate General Mengozzi).

[Delete last sentence of fifth paragraph: "It is considered that this case should be treated with caution." Add the following text: page 536]

The ECJ has subsequently confirmed that the following cumulative tests must be met for the *Hamburg Waste* exception to apply: First, the contract must be to establish co-operation with the aim of ensuring that a public task that all of the authorities have to perform is carried out; secondly, the contract is concluded exclusively by public entities, without the participation of a private party; thirdly, no private provider of services is placed in a position of advantage vis-à-vis competitors; and lastly, the co-operation is governed solely by considerations and requirements relating to the pursuit of objectives in the public interest.⁶⁴ᴬ

⁶⁴ᴬ *Azienda Sanitaria Locale di Lecce v Ordine degli Ingegneri della Provincia di Lecce* (C-159/11) [2013] 2 C.M.L.R. 17; *Piepenbrock Dienstleistungen GmbH & Co. KG v Kreis Düren* (C-386/11) [2014] 1 C.M.L.R. 1.

Application of a lesser regime.

[Add at the end of note 67: page 536]   15–012

See also, *R. (On the Application of Hoole and Co) v Legal Services Commission* [2011] EWHC 886 (Admin); *R. (On the Application of Harrow) v Legal Services Commission* [2011] EWHC 1087 (Admin); *R. (On the Application of All About Rights) v Legal Services Commission* [2011] EWHC 964 (Admin), and *R. (On the Application of Hossacks) v Legal Services Commission* [2012] EWCA Civ 1203.

[Add new second paragraph: page 536]

The 2014 Directives, and the Regulations that it is expected will implement them into UK law, will do away with the distinction between Part A and Part B services. Instead, those services listed under Annex XIV of Directive 2014/24 and under Annex XVII of Directive 2014/25 (which overlap significantly with those currently designated Part B services) will be subject to a discrete advertising and award regime.⁷⁰ᴬ This is referred to in the recitals to those Directives as the "light regime",⁷⁰ᴮ the details of which have been left to the Member States to decide.

⁷⁰ᴬ Directive 2014/24 arts 74–77; Directive 2014/25 arts 91–94.
⁷⁰ᴮ Directive 2014/24 Recital (28); Directive 2014/25 Recital (36).

Chapter Fifteen—Public Procurement

Procedures for the award of public contracts.

15–013 [Add to the end of note 73: page 537]

The discretionary grounds include grave misconduct in the course of a business or profession, which can include breach of contract, although the conduct must denote a wrongful intent or negligence of certain gravity: *Forposta SA v Poczta Polska SA* (C-465/11) unreported, December 13, 2012.

[Add to the end of note 74: page 537]

See also *Consorzio Stabile Libor Lavori Pubblici v Comune di Milano* (C-358/12), where an Italian law, requiring the annulment of a contract following the discovery that the economic operator in question owed €278 in social security payments, was held to be both proportionate and in pursuit of a legitimate objective.

15–014 [Add new fifth paragraph at end of text: page 540]

A further procedure has been introduced in the 2014 Directives. This is known as the "Innovation Partnership".[99A] While similar in many ways to the competitive dialogue process, the innovation partnership differs in that an economic operator will be chosen to develop a solution, rather than being chosen to implement a solution developed during the competitive dialogue.

[99A] Directive 2014/24 art.31; Directive 2014/25 art.49.

The bases for the award of public contracts.

15–015 [Add new second paragraph: page 540]

The 2014 Directives, once implemented, will abolish the separate "lowest price" basis. Instead, contracting authorities will be obliged to award contracts on the basis of the most economically advantageous tender, whichever procedure they adopt in doing so. However, the legislation allows the contracting authority to determine that, for the purposes of a particular procurement, the most economically advantageous tender will be the one offering the lowest price.[102A]

[102A] Directive 2014/24 art.67; Directive 2014/25 art.82.

Criteria and weightings.

15–016 [Add new note 102A to third sentence of first paragraph of 15–016: page 540]

The criteria used must be linked to the subject matter of the contract and include quality, price,[102A] technical merit, aesthetic and functional characteristics

¹⁰²ᴬ Use of fee percentages to measure price, on the assumption that costs would be the same for all bidders, was a manifest error: *Henry Brothers (Magherafelt) Ltd v Department of Education for Northern Ireland* [2011] NICA 59; [2012] B.L.R. 36.

[Add at the end of note 103: page 540]

In *European Commission v Kingdom of Netherlands* (C-368/10) [2013] All E.R. (EC) 804 the ECJ held that the requirement to use a specific eco-label breached Directive 2004/18 as the requirement should have referred to detailed underlying specifications for the product.

[Add at the end of note 106: page 541]

See also *Bundesdruckerei GmbH v Stadt Dortmund* (C-549/13) (incompatible with freedom to provide services to require payment of a minimum wage to workers not living in the contracting authority's Member State).

[Add at the end of note 107: page 541]

See also, *Easycoach Ltd v Department for Regional Development* [2012] NIQB 10 (on the objectivity of selection criteria).

[Add at the end of note 108: page 541]

The criteria must be clear, precise and unequivocal: *European Commission v Kingdom of Netherlands* (C-368/10) [2013] All E.R. (EC) 804. They must be such as would "allow all reasonably well informed and normally diligent tenderers to interpret them in the same way": *SIAC Construction Ltd v County Council of the County of Mayo* (C-19/00) [2001] E.C.R. I-7725. The standard of the "reasonably well informed and normally diligent tenderer" will be assessed objectively and not by reference to the aggrieved tenderer's actual interpretation: *Healthcare at Home Ltd v The Common Services Agency* [2014] 4 All E.R. 210; [2014] UKSC 49.

[Add at the end of note 109: page 541]

See also, *Evropaiki Dynamiki v European Commission* (T-39/08) unreported, December 8, 2011; *Evropaiki Dynamiki v Court of Justice of the European Union* (T-447/10), unreported, April 24, 2012.

[Add at the end of note 114: page 542]

See also, *Mears Limited v Leeds CC* [2011] Eu.L.Rev. 764.

Abnormally low tenders.

15–018 [Add at the end of note 119: page 543]

See also, *SAG ELV Slovensko a.s v úrad pre verejné obsarávanie* (C-599/10) [2012] 2 C.M.L.R. 36.

[Add to note 120: page 543]

Where the contracting authority does not suspect that a tender is abnormally low, it is under no duty to investigate it: *Resource (NI) Ltd v Ulster University* [2013] NIQB 64, [48].

[Add to first paragraph of 15–018 following: "When considering whether to reject an offer on the grounds that it is abnormally low; and (d) to take a decision as to whether to admit or reject the tenders.[120]": page 543]

A proper exchange of views between the contracting authority and the tenderer to enable the latter to demonstrate that its' tender is genuine is a fundamental requirement of the Directive.[120A]

[120A] *SAG ELV Slovensko a.s v úrad pre verejné obsarávanie* (C-599/10) [2012] 2 C.M.L.R. 36.

[Add at the end of note 121: page 543]

See also, *Amey LG Limited v The Scottish Minsters* [2012] CSOH 181.

[Add at the end of note 123: page 543]

See also, *SAG ELV Slovensko a.s v úrad pre verejné obsarávanie* (C-599/10) [2012] 2 C.M.L.R. 36.

[Add new third paragraph: page 544]

The 2014 Directives introduce specific circumstances in which the contracting authority must reject an abnormally low tender,[124A] namely where it is abnormally low by reason of the tenderer's failure to comply with its obligations under environmental, social or labour law.[124B]

[124A] Directive 2014/24 art.69(3); Directive 2014/25 art.84(3).
[124B] Directive 2014/24 art.18(2); Directive 2014/25 art.36(2).

Late tenders.

15–019 [Add to end of note 125: page 544]

See also, *R. (On the Application of Greenwich Community Law Centre) v Greenwich London BC* [2011] EWHC 3463 (Admin) (whether rejection of late bid was irrational or in manifest error); and *R (on the application of All*

About Rights Law Practice) v The Lord Chancellor [2013] EWHC 3461 (Admin) (not disproportionate to reject bid that was submitted incomplete).

[Add to text at end of paragraph: page 544]

However, contracting authorities may request additional documentation from a tenderer after the deadline for submission, provided that the submission of such documentation was not expressly made a prerequisite for consideration under the contract documents.[125A]

[125A] *Ministeriet for Forskning, Innovation og Videregaende Uddannelser v Manova A/S* (C-336/12) [2014] P.T.S.R. 254.

Award of a public contract.

[Add to end of note 126: page 544]

15–020

See also, *Rutledge Recruitment and Training Limited v Department for Employment and Learning* [2011] NIQB 61 and *Montpellier Estates Limited v Leeds CC* [2013] EWHC 166 (QB).

Material variations.

[Add to end of note 135: page 546]

15–022

These principles have been applied in a number of cases against the Legal Services Commission, including *R. (On the Application of Hoole) v The Legal Services Commission* [2011] EWHC 886 (Admin); *R. (On the Application of Harrow) v The Legal Services Commission* [2011] EWHC 1087 (Admin); *R. (On the Application of All About Rights) v The Legal Services Commission* [2011] EWHC 964 (Admin); and *R. (On the Application of Hossacks) v The Legal Services Commission* [2012] EWCA Civ 103. See also, *Clinton (t/a Oriel Training Services) v Department of Employment and Learning* [2012] NICA 48.

[Add at end of first paragraph of 15–022 following "On the other hand it is permissible (and sometimes necessary) to seek to clarify ambiguities in a tender.[135]": page 546]

It is not necessary to clarify imprecise tenders and to do so would risk breach of the principle of equal treatment.[135A]

[135A] *SAG ELV Slovensko a.s. v úrad pre verejné obsarávanie* (C-599/10) [2012] C.M.L.R. 36.

[Add at end of note 137: page 546]

Where a contracting authority agrees to a change in the contracting party or its composition (in the case of a consortium), the time limit for challenging

CHAPTER FIFTEEN—PUBLIC PROCUREMENT

that decision will run from the date that it is communicated to the aggrieved tenderer: *Idrodinamica Spurgo Velox srl v Acquedotto Pugliese SpA* (C-161/13) [2014] P.T.S.R. 935.

Electronic procurement.

15–025 [Add to note 154: page 548]

The 2014 Directives, and the Regulations implementing them, will make it compulsory to provide electronic access to tender documents: see Directive 2014/24 art.53; and Directive 2014/25 art.73. Further, Directive 2014/55 on electronic invoicing, published in May 2014, provides that contracting authorities must accept e-invoices that comply with a standard form.

Utilities.

15–026 [Insert new footnote within second sentence of first paragraph of 15–026: page 549]

Utilities are relevant persons carrying on activities in the fields of gas, electricity, water, transport services,[162A] postal services"

[162A] For an analysis of activities within the field of transport by rail, see *Alstom Transport v Eurostar International Limited* [2012] EWHC 28 (Ch).

Litigation.

15–027 [Add new note 173A: page 550]

The amendments to the 2006 Regulations came into force in relation to contract... procedures commencing on or after December 20, 2009.[173A]

[173A] See *Covanta Energy Ltd v Merseyside Waste Disposal Authority* (2013) 151 Con. L.R. 146; [2013] EWHC 2922 (TCC).

[Substitute note 175: page 551]

[175] (SI 2011/2053).

Breaches of the Regulations.

15–028 [Amend last two sentences of first paragraph of 15–028 and add new text: page 551]

A court will not interfere with an authority's exercise of its discretion unless it has committed a manifest error. A helpful summary of the principles is set out in *Lion Apparel v Firebuy*.[179] Although manifest error is frequently described as a case where an error has clearly been made, it appears that what is required is very similar to, if not the same as, the *Wednesbury*[179A] test of irrationality in domestic judicial review proceedings.[179B] A court will not undertake a comprehensive review or become embroiled in the merits of the tender evaluation process.[179C]

[179] [2007] EWHC 2179 (Ch), [27–39].
[179A] *Associated Provincial Picture Houses Limited v Wednesbury Corporation* [1948] 1 K.B. 223, CA.
[179B] See *BY Development Limited v Covent Garden Market Authority* [2012] EWHC 2546 (TCC).
[179C] See *BY Development Limited v Covent Garden Market Authority* [2012] EWHC 2546 (TCC) and *Shetland Line (1984) Limited v Scottish Ministers* [2012] CSOH 99.

[Add to note 182: page 552]

See also *Nordecon AS v Rahandusministeerium* (C-561/12) [2014] P.T.S.R. 343 (contracting authority may not negotiate with economic operator whose tender does not comply with mandatory requirements set out in the invitation to tender); and *Serco Belgium SA v European Commission* (T-644/13) (contracting authority not obliged to seek clarification where tender does not appear to comply with tendering specifications).

[Add to note 184: page 552]

See also, *Easycoach Limited v Department for Regional Development* [2012] NIQB 10 and *William Clinton trading as Oriel Training Services v Department for Employment and Learning* [2012] NICA 48.

Obtaining information before commencing a claim.

[In the second sentence of second paragraph of 15–029 add new note 186A as follows: page 553]:

15–029

Further information on the characteristics and relative advantages[186A] of the successful tenderer is available during the standstill period,

[186A] See fnn.187A and 187B, below.

[Add two new sentences to the end of the second paragraph of 15–029: page 553]

It is not necessary for the contracting authority to undertake a detailed comparative analysis of the successful and unsuccessful tenders as the provision of succinct comments on both tenders may be sufficient. However, the comments must be sufficiently precise to enable the unsuccessful bidder to ascertain the matters of fact and law on the basis of which the contracting authority rejected its offer and accepted that of the winning tenderer.[187A] It is not sufficient merely to provide the scores of both bidders as this does not enable the losing bidder to determine the specific reasons for the decision that the successful bid was better.[187B]

[187A] See *Evropaiki Dynamiki v European Commission* (C-561/10P) September 20, 2011; *Evropaiki Dynamiki v European Commission* (C-629/11P) October 4, 2012; and *Evropaiki Dynamiki v Court of Justice of the European Union* (T-447/10) October 17, 2012.
[187B] *Alfastar Benelux SA v Council of the European Union* (T-57/09) October 20, 2011.

Obtaining information during a claim.

15–030 [Add new text at end of the first paragraph: page 553]

Broad principles applicable to applications for early disclosure in procurement cases include:

(i) subject to issues of proportionality and confidentiality, the claimant ought to be provided promptly with the essential information and documentation relating to the evaluation process carried out, so that an informed view can be taken of its fairness and legality;

(ii) notwithstanding this, applications must be considered on their individual merits and a clear distinction may often be made between cases where a prima facie case has been made out by the claimant (but further information or documentation is required) and cases where the unsuccessful tenderer is aggrieved at the result but appears to have little or no grounds for disputing it;

(iii) applications must be tightly drawn and properly focused: information likely to be disclosable on an early application is that which demonstrates how the evaluation was actually performed and, therefore, why the claiming party lost; and

(iv) the court needs to balance the claimant's lack of knowledge and the need to guard against an application being used as a fishing exercise, designed to shore up a weak claim which will put the defendant to needless and unnecessary cost.[189A]

However, if sufficient information has been disclosed to enable the case to be pleaded and/or to demonstrate that there is a serious issue to be tried for the purposes of an application to set aside an automatic suspension,[189B] it may be difficult to obtain further documents prior to standard disclosure.[189C]

[189A] *Roche Diagnostics Limited v Mid Yorkshire Hospitals NHS Trust* [2013] EWHC 933 (TCC); *Pearson Driving Assessments Limited v The Minister for the Cabinet* [2013] EWHC 2082 (TCC); and *Covanta Energy Limited v*

Merseyside Waste Disposal Authority (2013) 151 Con. L.R. 135; [2013] EWHC 2964 (TCC).

[189B] (SI 2006/5) reg.47H.

[189C] *Pearson Driving Assessments Limited v The Minister for the Cabinet* [2013] EWHC 2082 (TCC) and *Covanta Energy Limited v Merseyside Waste Disposal Authority* (2013) 151 Con. L.R. 135; [2013] EWHC 2964 (TCC).

[Add at end of note 190: page 553]

In appropriate circumstances, a third-party disclosure order may be made against the winning tenderer: *NP Aerospace v Minstry of Defence* (unreported).

[Add at the end of note 192: page 554]

See also, *Healthcare at Home Limited v The Common Services Agency* [2011] CSOH 22.

Limitation.

[Add a new sentence to the first paragraph of 15–031 following: "It has been held that ... it is not necessary to have knowledge of potential loss or damage.[200]": page 555]

Nonetheless, suspicion will not be enough, although reasonable belief will suffice.[200A]

[200A] *Sita UK Limited v Greater Manchester Waste Disposal Authority* [2011] EWCA Civ 156; *Mermec UK Limited v Network Rail Infrastructure Limited* [2011] EWHC 1847 (TCC); and *Nationwide Gritting Services Limited v Scottish Ministers* [2013] CSOH 119.

[Amend second sentence of note 202 as follows: page 555]

A new limitation period, taking account of the decision in *Uniplex* [2010] 2 C.M.L.R. 47, was included in the Defence and Security Public Contracts Regulations reg.53,

[Add a new paragraph at the end of the third paragraph of 15–031: page 555]

The time limits have been strictly applied by the courts and the limitation period (now 30 days) is likely to start to run on the day on which the letter is received by the unsuccessful bidder, setting out the reasons why it lost, unless a factor such as illness of critical personnel prevented consideration of the letter.[204A]

[204A] *Mermec UK Limited v Network Rail Infrastructure Limited* [2011] EWHC 1847 (TCC) and *Turning Point Limited v Norfolk CC* [2012] EWHC

15–031

CHAPTER FIFTEEN—PUBLIC PROCUREMENT

2121 (TCC). See also, *Corelogic Limited v Bristol CC* [2013] EWHC 2088 (TCC) and *Montpellier Estates Limited v Leeds CC* [2013] EWHC 166 (QB).

15–032 [Add new sentence at the end of the second paragraph of 15–032: page 556]

It is necessary to show a good reason for an extension. Extensions may be granted where there are factors preventing issue of the claim which are beyond the control of the claimant, including illness or detention of the relevant personnel.[208A]

[208A] *Mermec UK Limited v Network Rail Infrastructure Limited* [2011] EWHC 1847 (TCC) and *Turning Point Limited v Norfolk CC* [2012] EWHC 2121 (TCC). See also, *Corelogic Limited v Bristol CC* [2013] EWHC 2088 (TCC) and *Montpellier Estates Limited v Leeds CC* [2013] EWHC 166 (QB). Compare the approach taken in Northern Ireland in *Traffic Signs and Equipment Limited v Department for Regional Development* [2010] NIQB 138; *Easycoach Limited v Department for Regional Development* [2012] NIQB 10; and *Henry Brothers (Magherafelt) Limited v Department of Education for Northern Ireland* [2011] NICA 59, [2012] B.L.R. 36.

Automatic suspension.

15–036 [Add new note 228A: page 559]

The courts have applied the *American Cyanamid* test in determining such applications by contracting authorities.[228A]

[228A] See *NATS (Services) Ltd v Gatwick Airport Ltd* [2014] EWHC 3133 (TCC).

[Add after first sentence of note 229: page 559]

See also, *Elekta Ltd v The Common Services Agency* [2011] CSOH 107; *Clinical Solutions International Limited v NHS 24* [2012] CSOH 10; *Shetland Line (1984) Limited v Scottish Ministers* [2012] CSOH 99; *Newcastle Upon Tyne Hospital NHS Foundation Trust v Newcastle Primary Care Trust* [2012] EWHC 2093 (QB); *Amey LG Limited v Scottish Ministers* [2012] CSOH 181; *Glasgow Rent Deposit & Support Scheme v Glasgow CC* [2012] CSOH 199; *Chigwell (Shepherds Bush) Limited v ASRA Greater London Housing Association Limited* [2012] EWHC 2746 (QB); and *Lowry Brothers Ltd v Northern Ireland Water Ltd* [2013] NIQB 23.

Where a declaration of ineffectiveness must not be made.

15–038 [Add note 232A: page 561]

Regulation 47L provides that, where ... effects of the contract should be maintained.[232A]

EUROPEAN COMMUNITY LEGISLATION

²³²ᴬ See also *Ministero dell'Interno v Fastweb SpA* (C-19/13), dealing with similar provisions under Directive 89/665 as amended by Directive 2007/66.

Judicial review.

[Add at the end of note 250: page 564]   **15–042**

See also, *R. (On the Application of Unison) v NHS Wiltshire Primary Care Trust* [2012] EWHC 624 (Admin) and *Traffic Signs and Equipment Limited and David Connolly v Department for Regional Development* [2012] NICA 18.

[Delete third paragraph of 15–042 and substitute with the following: page 565]

On July 1, 2013 new time limits came into effect regarding the commencement of applications for judicial review relating to decisions governed by the Public Contracts Regulations 2006.²⁵¹ In such cases, the claim form must be filed within the time limit in reg.47D(2). That limit is within 30 days of the date when the economic operator knew or ought to have known that grounds for starting the proceedings had arisen,²⁵¹ᴬ disregarding the rest of that regulation. The new time limit is not expressed to apply to applications for judicial review relating to decisions governed by the Utilities Contracts Regulations 2006 or the Defence and Security Public Contracts Regulations 2011.²⁵¹ᴮ

²⁵¹ CPR r.54.5(6).
²⁵¹ᴬ Where the decision being challenged has been taken in stages, the relevant date may not be that of the final stage in the decision-making process, where the earlier stages were intended to have legal effect: *R (Nash) v Barnet LBC* [2013] P.T.S.R. 1457; [2013] EWCA Civ 1004.
²⁵¹ᴮ Such judicial review challenges appear, therefore, to remain subject to the general judicial review time limit, as modified to take account of the decision in *Uniplex (UK) Limited v NHS Business Services Authority* (C-406/08) [2010] C.M.L.R. 47: see CPR r.54.5(1) and *R. (On the Application of Berky) v Newport CC* [2012] EWCA Civ 378.

Commission remedies.

[Amend references within the text of this paragraph and accompanying **15–043** notes 254 and 258: pages 565 and 566]

For "art.226 EC" substitute with "art.258 TFEU (ex art.226 EC)".

[Amend references within the text of this paragraph: pages 565 and 566]

For "art.228(2) EC" substitute with "art.260(2) TFEU (ex art.228(2) EC)".

CHAPTER FIFTEEN—PUBLIC PROCUREMENT

[Amend citation in first sentence of note 254 as follows: page 565]

Star Fruit v Commission (C-247/87) [1989] E.C.R. 291.

[Amend reference in first sentence of note 256 as follows: page 565]

For "(2010/C 177/01) art.83(2)" substitute with "([2012] O.J. L 265/1) art.160(3)"

2. COMMON LAW REMEDIES RELATING TO PUBLIC PROCUREMENT

Implied contract.

15–045 [Delete last sentence of note 269 and substitute with the following: page 568]

This view has been confirmed, albeit obiter, in *JBW Group Ltd v Ministry of Justice* [2012] EWCA Civ 8.

[Add to the text at the end of final paragraph of 15–045: page 568]

Where the Regulations do not apply because the contract is expressly excluded from the regime (as is the case for a services concession), the existence and content of an implied contract will depend on the common intention of the parties, whether it is necessary to imply terms for the purposes of efficacy and the terms of the tender documents.[269A]

[269A] See *JBW Group Ltd v Ministry of Justice* [2012] EWCA Civ 8 as to an obligation to consider the tender in good faith but that there was no common intention nor was it necessary to apply the detailed requirements of the Regulations. See also, *Montpellier Estates Limited v Leeds CC* [2013] EWHC 166 (QB); *Photo-Me International PLC v Network Rail Infrastructure Limited* [2011] EWHC 3168 (QB); and *Turning Point Limited v Norfolk CC* [2012] EWHC 2121 (TCC).

3. LOCAL GOVERNMENT PROCUREMENT LEGISLATION

15–046 [Add the following to the end of para.15-046: page 569]

The Public Services (Social Value) Act 2012 received Royal Assent on March 8, 2012. It applies to public services contracts, or services contracts together with the purchase or hire of goods or the carrying out of works,[274] to which the Regulations apply.[275] The Act also applies to framework agreements for which public services are likely to constitute the greater part by value of contracts,[276] but call-offs from framework agreements are not covered.[277] The operative provisions came into force on January 31, 2013.[278]

Where it applies, the Act requires authorities to consider how what is to be procured might improve the economic, social and environmental well-being of the "relevant area"[279], how it might act with a view to securing that improvement in conducting the process of procurement,[280] and whether to undertake consultation.[281] Consideration of social value is to take place at the "pre-procurement" planning stage[282], the intention being to ensure that social value is taken into account in setting the specification, tender rules and criteria.

The Localism Act 2011 received Royal Assent on November 15, 2011, as part of the Government's "Big Society" pledge to devolve more power to local communities. It includes the "Community Right to Challenge"[283]. This gives certain bodies (such as parish councils, employees and voluntary bodies)[284] the opportunity to submit an expression of interest[285] to a relevant authority (such as local or county councils)[286] to run a service or part of a service. Under s.81(1) of the Act, the relevant authority must consider the expression of interest. It must then either accept or reject it.[287] Where it accepts an expression of interest, it must run a procurement exercise for the service to which the expression of interest relates.[288] That procurement must be appropriate having regard to the value and nature of the contract.[289] The authority can only reject the expression of interest on designated grounds.[290] The provisions of the Act relating to the Community Right to Challenge came into force on June 27, 2012.[291]

Also of significance to public procurement are the provisions of the Localism Act 2011 relating to EU financial sanctions.[292] These give Government Ministers the ability to pass on part or all of a fine imposed by the European Court of Justice on the UK Government (further to EU infraction proceedings under art.260 of the Treaty on the Functioning of the European Union) to a designated public authority.[293] The designation procedure requires consultation with the authority in question and an order adopted by Parliament.[294] These provisions, which have given rise to much controversy, are designed to ensure that local and other public authorities are held to account for their part in EU infringements, including public procurement breaches. The provisions of the Act relating to EU sanctions came into force on May 31, 2012.[295]

[274] Public Services (Social Value) Act 2012 s.1(1)(a).
[275] Public Services (Social Value) Act 2012 s.1(13). The Act applies to England and Wales (s.4(4)), but has limited application in Wales where functions are devolved (see s.1(11)).
[276] Public Services (Social Value) Act 2012 s.1(1)(b).
[277] Public Services (Social Value) Act 2012 s.1(1)(a).
[278] Public Services (Social Value) Act 2012 (Commencement) Order 2012 (SI 2012/3173).
[279] Public Services (Social Value) Act 2012 s.1(3)(a). "Relevant area" is defined by s.1(4) as "the area consisting of the area or areas of the one or

more relevant authorities on whose behalf a public services contract is, or contracts based on a framework agreement are, intended to be made."

[280] Public Services (Social Value) Act 2012 s.1(3)(b).
[281] Public Services (Social Value) Act 2012 s.1(7).
[282] Public Services (Social Value) Act 2012 s.1(2).
[283] Localism Act 2011 s.81.
[284] Localism Act 2011 s.81(6).
[285] Localism Act 2011 s.81(4). See s.82 for timing of expressions of interest.
[286] Localism Act 2011 s.81(2).
[287] Localism Act 2011 s.83(1).
[288] Localism Act 2011 s.83(2). The relevant authority must consider whether acceptance of that expression of interest would promote or improve the social, economic or environmental well-being of the authority's area (s.83(8)).
[289] Localism Act 2011 s.83(3).
[290] See s.83(11) and Sch.1 to The Community Right to Challenge (Fire and Rescue Authorities and Rejection of Expressions of Interest) (England) Regulations 2012 (SI 2012/1647).
[291] Localism Act 2011 (Commencement No.6 and Transitional, Savings and Transitory Provisions) Order 2012 (SI 2012/1463).
[292] Localism Act 2011 Pt 2 ss.48–57.
[293] Localism Act 2011 s.48. See s.51 for the public authorities covered by this Part.
[294] Localism Act 2011 s.52.
[295] Localism Act 2011 (Commencement No.5 and Transitional, Savings and Transitory Provisions) Order 2012 (SI 2012/1008).

Chapter Sixteen

VARIOUS LEGISLATION

☐ 1.	Defective Premises Act 1972	16–001
2.	Control of Pollution Act 1974 and Environmental Protection Act 1990	16–008
☐ 3.	Limitation Act 1980 and Latent Damage Act 1986	16–013
☐ 4.	Building Act 1984 and the Building Regulations	16–025
5.	Insolvency Act 1986	16–032
	☐ (a) *Insolvency generally*	16–032
	(b) *Stay of proceedings—winding-up by the court*	16–035
	☐ (c) *Insolvency of contractor*	16–037
	(d) *Insolvency of employer*	16–046
6.	Party Wall, etc. Act 1996	16–047
	(a) *Party walls and party structures*	16–048
	(b) *Building Owners and Adjoining Owners*	16–053
	(c) *Serving of notices*	16–054
	(d) *The works permitted by the Act*	16–055
	☐ (e) *Dispute resolution under the Act*	16–059
	☐ (f) *Remedies*	16–063
	(g) *Rights and obligations incidental to the works*	16–066
	(h) *Capacity and liability of surveyors*	16–070
■ 7.	Anti-Social Behaviour Act 2003	16–071

1. DEFECTIVE PREMISES ACT 1972

The duty to build dwellings properly.

[Add to text after the sentence "It presumably applies…floors, doors, etc.": **16–002** pages 571–572]

In a block of flats a "dwelling" comprises each apartment, together with its balcony to which the occupier of the apartment had exclusive access for living. The common parts and basement car park do not constitute part of the "dwelling"[10A]

CHAPTER SIXTEEN—VARIOUS LEGISLATION

[10A] *Rendlesham Estates Plc v Barr Ltd* [2014] EWHC 3968 (TCC).

[Insert new paragraph after the sentence "Whilst the Act covers....to an existing dwelling": page 572]

The Act applies not merely to the "dwelling" but to "work for or in connection with the provision of a dwelling". Whilst this will be a question of fact in each case, the structural and common parts of a block of apartments fall within this description. Thus where two separate blocks of apartments were constructed by the same builder to the same specification and by the terms of the leases of each apartment, each occupier was liable to make payments to a management company responsible for the maintenance and repair of both blocks and each owner of an apartment had a right of access to the common parts of both blocks, then the structural and common parts provided in one block could be said to have been carried out in connection with the provision of an apartment in another block.[11A]

[11A] *Rendlesham Estates Plc v Barr Ltd* [2014] EWHC 3968 (TCC).

[Add following the sentence after "Where there are a number of defects ... fit for habitation.[13]": page 572]

In *Harrison v Shepherd Homes Ltd* it was held that defects in the foundations had rendered properties unfit for habitation notwithstanding the fact that the damage to the properties themselves was relatively minor.[13A]

[13A] [2011] EWHC 1811 (TCC). Ramsey J.: "Whilst I do not consider that the damage to the properties has rendered them unfit for habitation, on balance, I am persuaded that any significant defects in foundations are properly matters which could be said to give rise to a lack of fitness for habitation" (at [164]).

[Insert new paragraph after the sentence "Further, "if , when the work....have not then become patent."": page 572]

In *Rendlesham Estates Plc*[16A] the court set out helpful guidance as to the standard to be applied in considering whether a dwelling is "fit for habitation". Thus it must be capable of occupation for a reasonable time without risk to the health or safety of the occupants and without undue inconvenience or discomfort to the occupants. In applying these tests, the date of completion of the work is the relevant date for these purposes and if the dwelling would not be approved under the Building Regulations as fit for occupation it would probably not be "fit for habitation". The defects must be considered as a whole when applying this test and must be considered in the light of all the types of person who might reasonably be expected to occupy the dwelling including babies and those who suffer from common conditions such as asthma or hay fever. The fact that a defect can be remedied at relatively modest cost does not mean there is no breach of the duty under s.1 of the Act. Serious inconvenience, such as frequent breakdowns of a lift, may render a dwelling unfit for occupation. Furthermore, a risk of failure within the design life of the building of a structural element of the dwelling or the building of which the dwelling forms part, which exists at the date of completion, whether patent or latent, may make the dwelling unfit for occupation.

[16A] *Rendlesham Estates Plc v Barr Ltd* [2014] EWHC 3968 (TCC).

Persons owing the duty.

[Add to note 23: page 574]: **16–004**

See also, *Zenstrom v Fagot* [2013] EWHC 288 (TCC).

Measure of damages.

[Insert new paragraph: page 575] **16–007**

In *Rendlesham Estates Plc*[33A] the court rejected the suggestion that the damages for each apartment owner should be restricted to that owner's contribution to the management company of the apartment blocks of the cost of the necessary repairs. Instead each owner was entitled to the full cost of repairing the defect necessary to make his flat fit for occupation, including the cost of remedying the common parts where this was required to achieve that result. In order to avoid over-recovery of damages, the relevant judgment sum could only be enforced against the builder once and then on condition that damages were paid to the claimants' solicitor to hold the money for the benefit of the management company to enable it to carry out the necessary repairs.

[33A] *Rendlesham Estates Plc v Barr Ltd* [2014] EWHC 3968 (TCC).

3. LIMITATION ACT 1980 AND LATENT DAMAGE ACT 1986

[Add the following sentence to note 68 after: "Paragraph 5.1 of CPR **6–013** Practice Direction 7A...the claim is 'brought' on that earlier date for the purposes of the Limitation Act": page 579]

For circumstances in which a claim form was allegedly lost by the court, see *Page v Hewetts Solicitors (A Firm)* [2012] EWCA Civ 805.

CHAPTER SIXTEEN—VARIOUS LEGISLATION

Contract.

16–014 [Add to text after second sentence of first paragraph: page 579]

Nor does a mere procedural bar to the bringing of a claim prevent time from running in respect of the cause of action to which it relates.[73A]

[73A] *Sevcon Ltd v Lucas CAV Ltd* [1986] 1 W.L.R. 462, applied more recently in the context of a construction contract in *JJ Metcalfe v Dennison* (December 6, 2013) (unreported).

[Add the following sentence (in first paragraph of 16–014) after: "An acknowledgement by reference to extrinsic evidence.[75]": page 579]

However, the time limit may effectively be curtailed by a contractual limitation provision to the contrary.[75A]

[75A] *Inframatrix Investments Limited v Dean Construction Limited* [2012] EWCA Civ 64. Cf. *Elvanite Full Circle Limited v AMEC Earth & Environmental (UK) Limited* [2013] EWHC 1191 (TCC).

Tort.

16–015 [Add new second paragraph: page 582]

If the damage suffered is the incurrence of a contingent liability to a third party, there must be some additional and measurable loss suffered before time will begin to run. This will usually occur when the third party brings a claim, such that the contingent liability becomes an actual liability. However, additional loss has also been held to occur upon the incurrence of the contingent liability in two categories of case: those under the "damaged asset" rule and those under the "package of rights" rule.[88A] The "damaged asset" rule operates where the defendant's negligence causes an existing asset belonging the claimant to become encumbered, thereby reducing its value, e.g. by taking out a mortgage on a property.[88B] The "package of rights" rule operates where the value of benefits that the claimant should have received under a transaction is, by the defendant's negligence, diminished or extinguished, e.g. by losing title to a property.[88C] These categories are not mutually exclusive. For example, a sub-contractor who delivers defective works to a contractor may diminish the value of the contractor's package of rights under the sub-contract and also cause damage to a pre-existing asset belonging to the contractor, namely the contractor's interest in the benefit of the main contract.[88D]

[88A] *Axa Insurance v Akther & Darby* [2010] 1 W.L.R. 1662.
[88B] *Forster v Outred & Co* [1982] 1 W.L.R. 86. The asset damaged need not be physical: *Co-operative Group Ltd v Birse Developments Ltd (In Liquidation)* (2014) 153 Con. L.R. 103; [2014] EWHC 530 (TCC).

⁸⁸ᶜ *Bell v Peter Browne & Co* [1990] 2 QB 495. The transaction need not be between the claimant and the defendant: *DW Moore v Ferrier* [1988] 1 W.L.R. 267.

⁸⁸ᴰ *Co-operative Group Ltd v Birse Developments Ltd (In Liquidation)* (2014) 153 Con. L.R. 103; [2014] EWHC 530 (TCC).

[Add to note 93: page 583]

and *Gallagher v ACC Bank Plc* [2012] P.N.L.R. 29.

[Add to note 97: page 583]

See also, *Green v Eadie* [2012] Ch. 363; *Boycott v Perrins Guy Williams* [2011] EWHC 2969 (Ch).

Latent damage.

[Add to note 111: page 585]   16–017

; *Hunt v Optima (Cambridge) Ltd* [2014] P.N.L.R. 29, [2014] EWCA Civ 714.

Fraud, concealment or mistake.

[Add to note 114: page 585]   16–019

, upheld by the Supreme Court on this point, *Test Claimants in the FII Group Litigation v Revenue and Customs Comrs* [2012] 2 A.C. 337.

[Add to note 115: page 585]

In *Mortgage Express v Abensons Solicitors* [2012] EWHC 1000 (Ch) however, it was said that a limitation defence would only be lost where the party in default was aware of a fiduciary obligation owed which had been consciously breached.

Amendments.

[Add to note 138: page 588]   16–022

See also, *Nokia Corp v AU Optronics Corp* [2012] EWHC 731 (Ch); *Berezovsky v Abramovich* [2011] EWCA Civ 153.

Adjudication.

[Add new footnote 158A to text: pages 590–591]   16–024

An adjudicator's decision gives rise to an independent cause of action for payment of a sum decided to be due.¹⁵⁸ᴬ

CHAPTER SIXTEEN—VARIOUS LEGISLATION

[158A] *Aspect Contracts (Asbestos) Ltd v Higgins Construction plc* [2014] 1 W.L.R. 1220; [2013] EWCA Civ 1541, per Longmore L.J., at [3].

[Delete the fourth sentence of the paragraph, and add the following: page 591]

The current position is that an unsuccessful party has six years in which to challenge the correctness of an adjudicator's decision even if the underlying cause of action has become statute barred.[159]

[159] *Jim Ennis Construction Ltd v Premier Asphalt Ltd* [2009] EWHC 1906 (TCC), (2009) 125 Con. L.R. 141; *Aspect Contracts (Asbestos) Ltd v Higgins Construction plc* [2014] 1 W.L.R. 1220, [2013] EWCA Civ 1541. Cf. *Walker Construction (UK) Ltd v Quayside Homes Ltd* (2014) 153 Con. L.R. 26, [2014] EWCA Civ 93.

4. BUILDING ACT 1984 AND THE BUILDING REGULATIONS

16–025 [Add to note 163 after "SI 2000/2532": page 591]

(repealed by SI 2010/2215);

Contravention.

16–028 [Add to note 172 after "(SI 1998/3129 reg.13) and ahead of final closing bracket: page 593]

, now revoked in England and Wales by SI 2010/404, but still in force in Scotland and Northern Ireland.

[Add to note 175 after: "*Ashby v Ebdon* [1985] Ch. 394": page 593]

, followed in *Eaton v Natural England* [2012] EWHC 2401 (Admin).

Liability of local authority.

16–031 [Add new second paragraph: page 595]

Where a local authority exercises its powers under the Building Act 1984, it will be liable to compensate any person who has sustained damage by reason of the exercise of those powers, unless it relates to a matter where the person has been in default.[188A] "In default" will be construed narrowly; for the local authority to escape liability, it must prove that the person in question was in breach of an obligation arising under the 1984 Act.[188B]

[188A] See Building Act 1984 s.106(1).
[188B] *Manolete Partners plc v Hastings BC* [2014] B.L.R. 389; [2014] EWCA Civ 562.

INSOLVENCY ACT 1986

5. INSOLVENCY ACT 1986

(a) Insolvency generally

[At note 189, delete "and the Banking Act 2009" and replace with the following: page 596] **16–032**

, the Banking Act 2009, the Legislative Reform (Insolvency) (Miscellaneous Provisions) Order 2010 (SI 2010/18), the Insolvency (Amendment) Rules 2010 (SI 2010/686), and the Insolvency (Amendment) (No.2) Rules 2010 (SI 2010/734).

[Amend reference in note 190: page 596]

Professor L. Sealy and Profesor D. Millman, *Annotated Guide to the Insolvency Legislation 2014*, 17th edn (London: Sweet & Maxwell, 2014);

[After last sentence of para.16–032 add the following: page 596]

However, commercial sense and absence of intention to evade insolvency laws are highly relevant factors in the application of the anti-deprivation principle. The modern tendency is to uphold commercially justifiable contractual provisions challenged as contravening the principle.[193A]

[193A] *Belmont Park Investments Pty v BNY Corporate Trustee Services Ltd* [2011] UKSC 38.

(c) Insolvency of contractor

Mutual dealings.

[Add to note 229: page 600] **16–040**

See also *Westshield Ltd v Whitehouse* [2014] Bus. L.R. 268; [2013] EWHC 3576 (TCC) (in the context of a company voluntary arrangement); and *Alexander & Law Ltd v Coveside (21BPR) Ltd* (2014) 152 Con. L.R. 163, [2013] EWHC 3949 (TCC) (winding-up petition, still to be decided, does not preclude summary enforcement).

Forfeiture, lien and seizure clauses.

[Add to note 230: page 600] **16–041**

See also fn.193A, above.

6. PARTY WALL, ETC. ACT 1996

(e) Dispute resolution under the Act

16–059 [Add to note 299: page 610]

The parties may agree to some other form of dispute resolution procedure, which may preclude recourse to the statutory procedure: *Dillard v F&C Commercial Property Holdings Ltd* [2014] EWHC 1219 (QB).

(f) Remedies

(i) Challenging the award—appeal

16–063 [Add to note 316: page 613]

In *Freetown Ltd v Assethold Ltd* [2013] 1 W.L.R. 701, it was held that the 14-day time limit for an appeal against the award begins on the date of receipt or deemed receipt of the award as opposed to the date of posting.

(iii) Remedies where works are not carried out in accordance with the Act

16–065 [Add to note 329: page 614]

Where the contravention is minor, it may not be appropriate to grant a mandatory injunction: *Rashid v Sharif* [2014] EWCA Civ 377.

7. ANTI-SOCIAL BEHAVIOUR ACT 2003

Definitions

16–072 [Add after the sentence "A line of evergreens ... above the ground level": page 618]

Thus it was held that a local authority was entitled to conclude that four trees backing onto a garden were not a "high hedge" within the meaning of the Act on the basis that they were individual trees which did not form a line as required.[355A]

[355A] *R (on the application of Castelli) v Merton LBC* [2013] EWHC 602 (Admin).

Complaints procedure

16–073 [Add to note 357: page 618]

For guidance as to the steps to be taken by a local authority considering a complaint made pursuant to s.65, see *R (on the application of Pelling) v Newham LBC* [2011] EWHC 3265 (Admin).

CHAPTER SEVENTEEN

ARBITRATION

☐ 1.	Alternative methods of dispute resolution	17–001
2.	Arbitration	17–003
☐ 3.	What is an arbitration agreement?	17–004
☐ 4.	Jurisdiction of the arbitrator	17–010
■ 5.	Disqualification for bias	17–021
6.	Ousting the jurisdiction of the court	17–023
☐ 7.	The right to insist on arbitration	17–025
☐ 8.	Arbitration procedure	17–038
☐ 9.	Control by the court	17–051
☐ 10.	International arbitration under ICC Rules	17–070

1. ALTERNATIVE METHODS OF DISPUTE RESOLUTION

[Amend reference in note 2: page 621]   **17–001**

; R. Gaitskell, *Construction Dispute Resolution Handbook*, 2nd edn (London: ICE Publishing, 2011); S. Blake, J. Browne, and S. Sime, *The Jackson ADR Handbook*, 1st edn (Oxford: Oxford University Press, 2014).

3. WHAT IS AN ARBITRATION AGREEMENT?

The statutory definition.

[Add to note 19: page 624]   **17–004**

See also *Mi-Space (UK) Ltd v Lend Lease Construction (EMEA) Ltd* [2013] B.L.R. 600, [2013] EWHC 2011 (TCC); and *Kruppa v Benedetti* [2014] 2 All E.R. (Comm) 617, [2014] EWHC 1887 (Comm).

Whether the procedure is arbitration.

17–007 [At to the end of note 33: page 626]

A clause referring "any difference or dispute" to "an independent expert" did not amount to an arbitration agreement: *Wilky Property Holdings plc v London & Surrey Investments Limited* [2011] EWHC 2226 (Ch). A clause providing for a decision to be agreed to by one of two party-appointed appraisers and the arbitrator was not an arbitration clause under the Arbitration Act 1996 as the Act envisaged that the arbitrator alone would make the decision: *Turville Heath Inc v Chartis Insurance UK Ltd (formerly AIG UK Ltd)* [2012] EWHC 3019 (TCC).

[Add new third paragraph: page 626]

Where the prescribed process does not require the parties to submit finally to a binding arbitration, the clause in question cannot be an arbitration clause. This is the case where, for example, the prescribed process provides for a two-stage mechanism involving an arbitral tribunal followed by recourse to the courts.[33A]

[33A] *Kruppa v Benedetti* [2014] 2 All E.R. (Comm) 617; [2014] EWHC 1887 (Comm).

4. JURISDICTION OF THE ARBITRATOR

17–010 [At to end of note 57: page 629]

; *Cukurova Holding AS v Sonera Holding BV* [2014] UKPC 15.

[Add to the end of the first paragraph of note 58: page 629]

See also *Abuja International Hotels Ltd v Meridien SAS* [2012] 1 Lloyd's Rep. 461 (the arbitration agreement must be treated as a distinct agreement); and *Malhotra v Malhotra* [2012] EWHC 3020 (Comm) (disputes under arbitration clause not limited).

17–012 [Add after the first sentence of note 69: page 631]

A deed was not sufficiently clear to mean that a third party's rights under an exclusion clause in the deed were subject to an arbitration clause in the deed: *Fortress Value Recovery Fund I LCC v Blue Skye Special Opportunities Fund LP* [2013] EWCA Civ 367.

17–013 [Add at the end of note 93: page 633]

; *Abuja International Hotels Ltd v Meridien SAS* [2012] 1 Lloyd's Rep. 461, and *Malhotra v Malhotra* [2012] EWHC 3020 (Comm).

Fraud.

[Add to text after note 114: page 635]   17–016

Similarly, allegations of criminal conduct in the performance of the contract will not deprive the arbitrator of jurisdiction in respect of other matters, in the absence of clear words to that effect.[114A]

[114A] *Interprods Ltd v De La Rue International Ltd* [2014] 1 Lloyd's Rep. 540 ; [2014] EWHC 68 (Comm).

Challenge to arbitrator's jurisdiction.

[Add to note 123: page 637]   17–017

; *Habas Sinai Ve Tibbi Gazlar Istihsal Endustrisi v VSC Steel Co Ltd* [2014] 1 Lloyd's Rep. 479, [2013] EWHC 4071 (Comm).

[Add to note 126: page 637]   17–018

However, once engaged, the provisions of s.72 are likely to be construed with "at least a degree of generosity": *London Steam Ship Owners Mutual Insurance Association Ltd v Spain (The Prestige)* (2013) 150 Con. L.R. 181, [2013] EWHC 2840 (Comm).

Appointment of judges as arbitrators.

[Amend reference in note 135: page 638]   17–020

, see s.18 of the *TCC Guide*, 2nd edn, 3rd revision (effective from March 3, 2014).

5. DISQUALIFICATION FOR BIAS

[Add to end of note 140: page 639]   17–021

applied in *Nestor Maritime SA v Sea Anchor Shipping Co Ltd* [2012] 2 Lloyd's Rep. 144.

[Add a new sentence to para.17–021 after: "Such bias was a species of serious irregularity which caused substantial injustice within the meaning of s.68 of the 1996 Act.[147]": page 640]

However, where an arbitrator had previously been instructed by solicitors acting for one party in the dispute on a matter wholly unrelated to the arbitration, this was not sufficient to lead a fair-minded and informed observer to conclude there was a possibility of bias.[147A]

[147A] *A v B* [2011] 2 Lloyd's Rep. 591.

CHAPTER SEVENTEEN—ARBITRATION

7. THE RIGHT TO INSIST ON ARBITRATION

A step in the proceedings.

17–029 [Add new sentence to the end of note 190: page 645]

A consent order made by the parties following a case management conference was an unequivocal step taken to answer the substantive claim: *Nokia Corp v HTC Corp* [2012] EWHC 3199 (Pat).

Valid arbitration agreement.

17–030 [Add new sentence at end of 17–030: page 645]

The court may also grant an injunction to prevent relevant proceedings.[191A]

[191A] *AES Ust-Kamenogorsk Hydropower Plant LLP v Ust-Kamenogorsk Hydropower Plant JSC* [2013] UKSC 35; *Excalibur Ventures LLC v Texas Keystone Inc* [2012] 1 All E.R. (Comm) 933.

No dispute.

17–031 [Add to note 198: page 646]

; *Laker Vent Engineering Ltd v Jacobs E&C Ltd* (2014) 154 Con. L.R. 77, [2014] EWHC 1058 (TCC). Where there has been a breach of the arbitration agreement but that breach is found not to be repudiatory, this will support the conclusion that the agreement remains operative: *BDMS Ltd v Rafael Advanced Defence Systems* [2014] 1 Lloyd's Rep. 576, [2014] EWHC 451 (Comm), [59].

8. ARBITRATION PROCEDURE

17–039 [Add to note 225: page 650]

However, any later agreement which purports to agree with any such matter may be precluded by the language of the arbitration clause, as s.34 is not a mandatory provision: *Secretary of State for Defence v Turner Estate Solutions Ltd* [2014] EWHC 244 (TCC), [83]–[97].

Section 42 of the 1996 Act.

17–040 [Add to note 239: page 651]

That intervention is at the court's discretion and not a "rubber-stamping exercise" was emphasised in *Patley Wood Farm LLP v Brake* [2013] EWHC 4035 (Ch).

Interim awards.

[Add new sentence to the end of note 252: page 653] 17–043

Certain appeal interim awards took effect as awards under s.47 of the 1996 Act and were final and binding: *Sucafina SA v Rotenberg* [2011] EWHC 901 (Comm), upheld on this at [2012] EWCA Civ 637.

9. CONTROL BY THE COURT

Extension of time.

[Substitute second sentence of note 300: page 659] 17–053

See also, *Nikola Rotenberg v Sucafina SA* [2012] EWCA Civ 637; *Buyuk Camlica Shipping Trading and Industry Co Inc v Progress Bulk Carriers Limited* [2010] EWHC 442 (Comm).

[Add new sentence to end of note 303: page 660]

A party who commenced an arbitration out of time and did not have a copy of the document which contained the arbitration clause with a time-bar but had requested the document from the other party, should have foreseen that the other party might fail to provide the document or that there might be a time-bar: *Perca Shipping Ltd v Cargill Inc* [2012] EWHC 3759.

Injunction to restrain arbitration proceedings.

[Substitute note 319: page 662] 17–056

See the discussion of the circumstances in which injunction may be granted under the 1996 Act in D. St John Sutton, *Russell on Arbitration*, 23rd edn (London: Sweet & Maxwell, 2007), Ch.7. An injunction was given to prevent a party from proceeding with a foreign arbitration where it was strongly arguable that the defendants were not parties to the arbitration agreement, and did not have a connection with the seat of the attempted arbitration or the arbitral body. There was power to grant the injunction under s.37 of the Senior Courts Act 1981 where s.9 of the 1996 Act did not apply: *Excalibur Ventures LLC v Texas Keystone* [2011] 2 C.L.C. 338.

Serious irregularity.

[Following the second sentence of note 326, add: page 662] 17–057

See also *Nestor Maritime SA v Sea Anchor Shipping Co. Ltd* [2012] 2 Lloyd's Rep. 144.

CHAPTER SEVENTEEN—ARBITRATION

[Add new sentence to the end of note 327: page 663]

There was no duty to alert a party to a possible argument favourable to that party based on a case not cited in a documents only arbitration: *ED&F Man Sugar Ltd v Belmont Shipping Ltd* [2012] 1 All E.R. (Comm) 962. It will be rare for a case management decision, such as the refusal of an adjournment, to give rise to a serious irregularity: *Brake v Patley Wood Farm LLP* [2014] EWHC 1439 (Ch).

[Add to the end of note 328: page 663]

Abuja International Hotels Ltd v Meridien SAS [2012] 1 Lloyd's Rep. 461.

[Substitute second sentence of note 329: page 663]

See also *Abuja International Hotels Ltd v Meridien SAS* [2012] 1 Lloyd's Rep. 461.

[Add to the end of note 329: page 663]

Where a tribunal dealt with an issue succinctly, in a single sentence and in a composite way, the issue had been sufficiently dealt with: *Petrochemical Industries Co (KSC) v Dow Chemical Co* [2012] Lloyd's Rep. 691.

[Add at the end of note 331: page 664]

Where there was fraud but, if the true position had been disclosed to the arbitrators, it would not have affected the result of the arbitration, it had not been shown either that the award was obtained by fraud or that it had caused substantial injustice: *Chantiers de L'Atlantique SA v Gaztransport & Technigaz SAS* [2011] EWHC 3383 (Comm).

Interim injunction.

17–061 [Add to end of text: page 667]

Similarly, the fact that the seat of the arbitration is inside England and Wales does not preclude recourse to the courts of some other jurisdiction where the agreement allows it.[351A]

[351A] *U&M Mining Zambia Ltd v Konkola Copper Mines plc* [2013] 2 Lloyd's Rep. 218; [2013] EWHC 260 (Comm).

Appeals to the High Court.

17–064 [Add new second sentence to note 363: page 668]

Questions of fact could not be dressed up as questions of law: *Geogas SA v Tramond Gas Ltd* [1993] 1 Lloyd's Rep. 215, 231–232, CA; *Demco Investments & Commercial SA v SE Banken Forsakring Holding Aktiebolag* [2005]

2 Lloyd's Rep. 650; *London Underground Ltd v Citylink Telecommunications Ltd* [2007] EWHC 1749 (TCC); and *House of Fraser Ltd v Scottish Widows Plc* [2011] EWHC 2800 (Ch), [25].

[Add to note 372: page 670] 17–065

See also *Morris Homes (West Midlands) Ltd v Keay* (2013) 152 Con. L.R. 105; [2013] EWHC 932 (TCC).

Leave to appeal is granted sparingly.

[Add at the end of note 384: page 671] 17–066

See also *HMV UK v Propinvest Friar Limited Partnership* [2012] 1 Lloyd's Rep. 416, [35–38], CA.

[Add after fourth sentence of para.17–066: "'Does it appear upon perusal of the award either that the arbitrator misdirected himself in law or that his decision was such that no reasonable arbitrator could reach?'[384"]: page 671]

Alternatively, was the decision "a major intellectual aberration?"[384A]

[384A] *Braes of Doune Wind Farm (Scotland) Ltd v Alfred McAlpine Business Services Ltd* [2008] 1 Lloyds Rep. 608, [31] as approved in *HMV UK v Propinvest Friar Limited Partnership* [2012] 1 Lloyd's Rep. 416, [8], CA.

10. INTERNATIONAL ARBITRATION UNDER ICC RULES

Proper law.

[Amend reference in note 469: page 679] 17–075

see generally, Lord Collins of Mapesbury et al., *Dicey, Morris and Collins* on *The Conflict of Laws*, 15th edn (London: Sweet & Maxwell, 2012).

CHAPTER EIGHTEEN

THE HOUSING GRANTS, CONSTRUCTION AND REGENERATION ACT 1996

	1.	Introduction	18–001
☐	2.	Overview	18–007
	3.	Adjudication under statute	18–015
■	4.	The approach to adjudication	18–018
☐	5.	Adjudication procedure	18–020
☐	6.	Enforcement of the adjudicator's decision	18–030
☐	7.	Grounds for opposing enforcement	18–035
	8.	Challenging an adjudicator's decision	18–053
■	9.	Payment	18–056
	10.	Right to suspend performance	18–063
	11.	Pay-when-paid clauses	18–065

2. OVERVIEW

Construction contracts.

[Add new sentence to text after note 33: page 689] **18–009**

A collateral warranty may, depending on its terms, be a construction contract under the Act.[33A]

[33A] *Parkwood Leisure v Laing O'Rourke Wales and West Ltd* (2013) 150 Con. L.R. 93; [2013] EWHC 2665 (TCC).

Excluded construction contracts.

[Add new sentence to the end of note 65: page 692] **18–012**

In determining whether there was an "intention to occupy", the date of formation of the contract was an important, but not an overriding factor.

CHAPTER EIGHTEEN—HOUSING GRANTS, CONSTRUCTION AND REGENERATION ACT 1996

The basis for the residential occupier exception was also questioned: *Westfields Construction Ltd v Lewis* [2013] B.L.R. 223.

4. THE APPROACH TO ADJUDICATION

Approach of the courts.

18–018 [Add new sentence to the end of note 99: page 697]

See also *Abbas (t/a A H Design) v Rotary (International) Ltd* [2012] NIQB 41.

5. ADJUDICATION PROCEDURE

Right to refer at any time.

18–020 [Add new first sentence at beginning of paragraph: page 698]

Under the statutory scheme, a dispute is referred to adjudication by means of a notice of adjudication,[102A] and it is upon the issue of that notice that proceedings are commenced.[102B]

[102A] See para.1 of the Scheme: see also, App.B below.
[102B] *University of Brighton v Dovehouse Interiors Ltd* (2014) 153 Con. L.R. 147; [2014] EWHC 940 (TCC), where the date of commencement determined the applicability of a conclusive evidence clause.

[Add new sentence to the end of note 105: page 698]

See also *NAP Anglia Ltd v Sun-Land Development Co Ltd* [2012] B.L.R. 110.

[Add new sentence following: "A party can give a valid notice of adjudication...brought to an end.": page 698]

The service of an adjudication notice and the appointment of an adjudicator did not prevent a party from commencing a new adjudication concerning the same dispute. It was only by a referral notice that the adjudicator had jurisdiction but even after a referral a party could omit a head of claim and reserve it for later adjudication.[107A]

[107A] *Lanes Group plc v Galliford Try Infrastructure Ltd (t/a Galliford Try Rail)* [2012] B.L.R. 121, CA.

[Add new second paragraph: page 699]

The right to refer under the Scheme is not, without more, an obligation to refer disputes to adjudication. On that basis, it has been held that the right to refer is not (for the purposes of s.1(4) of the Contracts (Rights of Third Parties) Act 1999) a term to which any express rights enjoyed by a third party under the contract are subject. As a result and absent express provision to the contrary, a third party will be neither obliged nor entitled to enforce its rights under the contract by referring a dispute to adjudication.[110A]

[110A] *Hurley Palmer Flatt Ltd v Barclays Bank plc* [2014] EWHC 3042 (TCC).

Appointment of adjudicator.

[Add new sentence after: "In practice this timetable is usually achieved. The appointment of the adjudicator is frequently carried out by an adjudicator nominating body.[111]": page 699] **18–021**

A procedure permitting one party to nominate the adjudicator is not in compliance with the Act.[111A]

[111A] *Sprunt Ltd v Camden LBC* [2012] B.L.R. 83, [45–51].

Decisions as to jurisdiction.

[Add new note 117A at end of sentence: "An adjudicator's jurisdiction is derived from the terms of their appointment as agreed by the parties,[117] and is accordingly a question of construction to which ordinary principles apply.": page 700] **18–022**

[117A] Where an unrepresented party paid one half of the adjudicator's fee and sent an email seeking guidance on the adjudication procedure, this was not sufficient to amount to submission to the jurisdiction of the adjudicator by that party: *Clark Electrical v JMD Developments (UK) Ltd* [2012] B.L.R. 546.

[Insert new final sentence into the second paragraph of 18–022: page 700]

The parties can of course provide that the adjudicator has the ability to decide his own jurisdiction.[122A]

[122A] *WSP Cel Ltd v Dalkia Utilities Services Plc* [2012] EWHC 2428 (TCC).

Adjudicator's fees and expenses.

[Add new sentence at the end of note 155: page 705] **18–029**

That decision was approved in *Systech International Ltd v PC Harrington Contractors Ltd* [2013] 2 All E.R. 69, [44].

CHAPTER EIGHTEEN—HOUSING GRANTS, CONSTRUCTION AND REGENERATION ACT 1996

[Add a new paragraph after the second paragraph of 18–029: page 705]

Where an adjudication decision is not enforceable due to a breach of natural justice or lack of jurisdiction the adjudicator may not be entitled to recover their fees.[158A]

[158A] *Systech International Ltd v PC Harrington Contractors Ltd* [2013] 2 All E.R. 69.

6. ENFORCEMENT OF THE ADJUDICATOR'S DECISION

Implied obligation to comply.

18–032 [Add new note 166A: page 707]

There is generally no ability to set-off sums from sums which are to be paid under an adjudicator's decision.[166A]

[166A] For a helpful summary of the cases on set-off against adjudicator's decisions, see *Squibb Group Ltd v Vertase FLI Ltd* [2012] B.L.R. 408, [19–27] and *Beck Interiors Ltd v Classic Decorative Finishing Ltd* [2012] EWHC 1956 (TCC).

Summary judgment.

18–033 [Add new sentence following: "Essentially for the policy reasons first...and for the service of evidence in opposition to a summary judgment application."[170"]: page 707]

The winding-up jurisdiction was not and never had been the proper forum for the determination of disputed claims. Where there was a bona fide defence of substance as to jurisdiction in relation to a debt that was not a judgment debt, it would not be just and equitable to make a winding-up order.[170A]

[170A] *Towsey (t/a Towsey Plastering Contractors) v Highgrove Homes Ltd* [2013] B.L.R. 45.

Stay of execution.

18–034 [Add to note 177: page 708]

; *Alexander & Law Ltd v Coveside (21BPR) Ltd* (2014) 152 Con. L.R. 163; [2013] EWHC 3949 (TCC).

[Add new sentence at end of 18–034: page 709]

The Court may grant a partial stay of execution where the successful party would be unable to pay part of the sum awarded.[184A]

¹⁸⁴ᴬ See *Nap Anglia Ltd v Sun-Land Development Co. Ltd* [2012] B.L.R. 110 and *Tate Building Services Ltd v B & M McHugh Ltd* [2014] EWHC 2971 (TCC).

7. GROUNDS FOR OPPOSING ENFORCEMENT

No dispute.

[Substitute fifth sentence of paragraph: page 713]   **18–042**

A referring party must refer a single dispute "arising out of a single contract",²¹⁷ but may make as many distinct referrals as it wishes.²¹⁷ᴬ

²¹⁷ᴬ *Willmott Dixon Housing Ltd v Newlon Housing Trust* (2013) 147 Con. L.R. 194; [2013] EWHC 798 (TCC).

[Add at beginning of note 217: page 713]

See, for example, *Viridis UK Ltd v Mulalley & Co Ltd* [2014] EWHC 268 (TCC).

[Add to end of note 217: page 713]

Whether a party is limited to a single dispute was doubted in *Willmott Dixon Housing Ltd v Newlon Housing Trust* (2013) 147 Con. L.R. 194; [2013] EWHC 798 (TCC), [75] and [77].

Disputes "under the contract".

[Substitute last sentence of second paragraph: page 714]   **18–043**

There is conflicting authority as to whether the reasoning in the decision in *Fiona Trust* applies to adjudication clauses.²²³ In any case, the contract must be read as a whole and the wording of other dispute resolution provisions may serve to narrow the scope of an adjudicator's jurisdiction.²²³ᴬ

²²³ᴬ *Hillcrest Homes Ltd v Beresford and Curbishley Ltd* (2014) 153 Con. L.R. 179; [2014] EWHC 280 (TCC).

[Add to note 223: page 714]

; cf. *Hillcrest Homes Ltd v Beresford and Curbishley Ltd* (2014) 153 Con. L.R. 179; [2014] EWHC 280 (TCC).

CHAPTER EIGHTEEN—HOUSING GRANTS, CONSTRUCTION AND REGENERATION ACT 1996

The existence of a dispute.

18–045 [Add at the end of note 242: page 717]

A dispute may crystallise over an interim valuation of work even where the sum is not yet due: *Working Environments Ltd v Greencoat Construction Ltd* [2012] B.L.R. 309.

No jurisdiction over an issue.

18–046 [Add new note 248A: page 718]

If the adjudicator purports to decide matters which, on a true construction of the referral documentation,... and will not be enforced.[248A]

[248A] It is the Notice of Adjudication that defines the scope of the referral, the Referral Notice only becoming relevant to that question where the Notice of Adjudication is ambiguous: *JG Walker Groundworks Ltd v Priory Homes (East) Ltd* [2013] EWHC 3723 (TCC).

[Add a new sentence at the end of note 252: page 718]

See also, *Vertase FLI Ltd v Squibb Group Ltd* [2012] EWHC 3194 (TCC) and *Carillion Construction Ltd v Smith* (2011) 141 Con. L.R. 117.

[Add a new sentence at the end of note 253: page 719]

If one disputed claim could not be decided without deciding all, or parts of, another disputed claim, that pointed to there being only one dispute: *Witney Town Council v Beam Construction (Cheltenham) Ltd* [2011] B.L.R. 707.

[Add new sentence after: "Where the parties had not given such consent, the adjudicator had no jurisdiction and its decision could not be enforced.[253]": page 719]

Where a single dispute was referred but additional questions were dealt with by oversight or error, any decision on those additional questions could be severed providing that the reasoning on that question did not form an integral part of the whole decision.[253A]

[253A] *Lidl UK GmbH v RG Carter Colchester Ltd* [2012] EWHC 3138 (TCC).

Mistaken answer to an issue.

18–048 [Add new sentence at the end of the paragraph: page 719]

An adjudicator might act in excess of jurisdiction if he went off on "a frolic of his own" and decided an issue using a method of assessment which had not been proposed by either party.[257A]

²⁵⁷ᴬ *Herbosch-Kiere-Marine Contractors Ltd v Dover Harbour Board* [2012] B.L.R. 177.

Breach of natural justice.

[Add to note 270: page 721]

18–050

; *ABB Ltd v BAM Nuttall Ltd* (2013) 149 Con. L.R. 172; [2013] EWHC 1983 (TCC).

[Insert after: "...where an adjudicator simply failed to consider relevant information which had been placed before him;²⁷⁶": page 721]

where an adjudicator decided an issue using a method of assessment proposed by neither party and which had not been put to either party;²⁷⁶ᴬ

²⁷⁶ᴬ *Herbosch-Kiere-Marine Contractors Ltd v Dover Harbour Board* [2012] B.L.R. 177. There was no breach of natural justice where an adjudicator used his own methodology to calculate target cost without putting it to the parties if the calculation was derived from figures and other findings based on the parties' submissions: *Hyder Consulting (UK) Ltd v Carillion Construction Ltd* 138 Con. L.R. 212. Where an adjudicator applied one of the parties' methods of assessment to derive a figure and also adjusted the project manager's figure but then split the difference between the two figures, this did not amount to a breach of natural justice: *Arcadis UK Ltd v May and Baker Ltd (t/a Sanofi)* [2013] EWHC 87 (TCC).

[Add new second sentence to note 277: page 721]

See also *Highlands and Islands Airports Ltd v Shetland Islands Council* [2012] CSOH 12 (Court of Session, Outer House).

[Add a new sentence at the end of note 281: page 722]

See also, *Berry Piling Systems Ltd v Sheer Projects Ltd* (2012) Con. L.R. 225 where the adjudicator's additional reason, not put to the parties, did not affect the decision.

[Add a new sentence at the end of note 283: page 722]

Severability of parts of a decision has been held to be possible where part of the decision falls outside the adjudicator's decision: *Working Environments Ltd v Greencoat Construction Ltd* [2012] B.L.R. 309 and *Lidl UK GmbH v RG Carter Colchester Ltd* [2012] EWHC 3138 (TCC).

CHAPTER EIGHTEEN—HOUSING GRANTS, CONSTRUCTION AND REGENERATION ACT 1996

Breach of human rights.

18–051 [Add a new sentence at the end of note 287: page 723]

See also, *Whyte and Mackay Limited v Blyth & Blyth* [2013] CSOH 54, [63] where art.6 rights were held not to be infringed even after taking into account the application of art.6 rights to interim proceedings as decided in *Micaleff* v *Malta* (2010) 50 EHRR 57, [74–89].

[Add a new sentence at the end of para.18–051: page 723]

In exceptional cases, enforcing the decision of an adjudicator might be a disproportionate interference with the right to peaceful enjoyment of possessions under art.1 of the First Protocol to the Convention.[287A]

[287A] *Whyte and Mackay Limited v Blyth & Blyth* [2013] CSOH 54, [46–47].

9. PAYMENT

Withholding notices under the unamended Act.

18–060 [Add new second sentence to note 342: page 730]

See also *Squibb Group Ltd v Vertase FLI Ltd* [2012] B.L.R. 408.

[Add new sentence at the end of note 343: page 730]

See also *R&C Electrical Engineers Ltd v Shaylor Construction Ltd* [2012] B.L.R. 373.

CHAPTER NINETEEN

LITIGATION

☐ 1.	Introduction	19–001
2.	Features of construction contract litigation	19–002
3.	Pre-action protocol and mediation	19–003
	☐ (a) *Pre-action protocol for construction and engineering disputes*	19–003
	☐ (b) *Mediation*	19–008
☐ 4.	Technology and Construction Court Judges (formerly Official Referees)	19–011
☐ 5.	County court jurisdiction	19–014
☐ 6.	Statements of case	19–017
7.	Particular pleadings	19–022
8.	Contractor's "claims"	19–028
9.	Scott Schedule	19–031
10.	Preparation for trial	19–041
	☐ (a) *Preparation of evidence*	19–042
	■ (b) *Disclosure and inspection of documents*	19–049
	■ (c) *Preliminary point*	19–052
	(d) *Plans, photographs, models, mechanically recorded evidence*	19–053
	■ (e) *View*	19–054
	☐ (f) *Settlement of actions*	19–056
☐ 11.	Set-off and counterclaim	19–057
☐ 12.	Summary judgment and interim payment	19–065
☐ 13.	Interest	19–072
☐ 14.	Costs	19–077
	■ (a) *Costs are discretionary*	19–080
	(b) *Summary assessment of costs*	19–084
	(c) *Notice to admit facts*	19–085

CHAPTER NINETEEN—LITIGATION

  (d) *Payment into court and offers to settle* 19–086
☐ (e) *Written offers* 19–087
■ (f) *Costs where there are cross-claims* 19–089

1. INTRODUCTION

19–001 [Change reference in note from "Civil Procedure 2011" to: page 738]

Civil Procedure 2014

For a further account, see *Civil Procedure 2014*, "The White Book" (London: Sweet & Maxwell, 2014) and S. Sime and D. French, *Blackstone's Civil Practice 2014* (Oxford: OUP, 2014).

[At sub-paragraph (b) remove "79 Parts" and substitute with: page 738]

86 Parts

[At sub-paragraph (c), after "justly" add: page 738]

and, as of April 1, 2013, at a proportionate cost.

[At sub-paragraph (f) substitute the reference to "Civil Procedure 2011" with: page 739]

Civil Procedure 2014

[Add new sub-paragraph (g): page 739]

  (g) A number of significant changes to the CPR (commonly referred to as the "Jackson Reforms") have also recently been implemented, with most of the reforms taking effect from April 1, 2013. Amongst other things, the reforms make important revisions to Pt 1 (Overriding Objective), Pt 3 (Case Management), Pt 31 (Disclosure), Pt 35 (Expert Evidence), Pt 36 (Offers to Settle), and Pts 43–48 (Costs), and can be found in a Special Supplement of the Civil Procedure 2013.

3. PRE-ACTION PROTOCOL AND MEDIATION

(a) Pre-action protocol for construction and engineering disputes

Introduction

19–003 [At note 11, substitute reference to "Civil Procedure 2011" with: page 740]

Civil Procedure 2014

[On the third line of the paragraph substitute reference to "Civil Procedure 2011" with: page 740]

Civil Procedure 2014

[On the fourth line of the paragraph substitute reference to "2nd edn" with: page 740]

3rd edn

[Substitute note 12: page 740]

Now in its Third Revision.

[Substitute second sentence of note 13: page 740]

The TCC Guide was revised with effect from March 3, 2014; the version set out in *Civil Procedure 2014*, Vol.2, at paras 2C-35 to 2C-167, is the superseded second revision.

Essential Elements of the Protocol

[Add to note 22: page 742]   **19–005**

See also, *Higginson Securities (Developments) Ltd v Hodson* [2012] EWHC 1052 (TCC), in which Akenhead J. suggested that whilst a meeting was not described as absolutely mandatory by the protocol, the default position was that a meeting ought to take place unless there was a reasonably good reason for it not to.

Compliance with the Protocol

[Delete the final sentence of note 23 and replace with the following: page **19–006** 742]

These recommendations did not ultimately form part of the implemented reforms, however (although it is to be noted that the last sentence of para.2.1.3 of the Guide appears to envisage an application to the court before the issue of proceedings; see also para.4.1.5).

[Add to note 25: page 742]

Cf. *Higginson Securities*, fn.22 above.

[Amend third sentence of paragraph: pages 742–743]

Where, however, the court is satisfied that a party has failed to comply ... proceedings until the relevant steps have been taken,[27] make an adverse costs order,[28] or impose such other conditions as it considers appropriate.[28A]

CHAPTER NINETEEN—LITIGATION

²⁸ᴬ The Guide, para.2.6.1.

[At note 28 substitute the reference to "r.44.3(v)(a)" with: page 743]

r.44.2(5)(a) (previously r.44.3(5)(a))

[Substitute last sentence of note 28: page 743]

For cases where the court ordered costs on an indemnity basis, see *Thomas Construction Ltd v Hyland* (2011) C.I.L.L. 1748, TCC; and *Sainsbury's Supermarkets Ltd v Condek Holdings Ltd (formerly Condek Ltd)* [2014] B.L.R. 574, [2014] EWHC 2016 (TCC).

[Add text at end of paragraph: page 743]

Where a claimant fails to serve proceedings in time, the fact that the claimant was attempting to comply with the pre-action protocol does not justify that failure and will seldom (if ever) support an application for an extension of time.²⁹ᴬ

²⁹ᴬ *Lincolnshire CC v Mouchel Business Services Ltd* [2014] B.L.R. 347; [2014] EWHC 352 (TCC).

Costs of compliance with the Protocol

19–007 [At note 32 delete the final sentence: page 743]

(b) Mediation

(i) Introduction

19–008 [Add to note 40: page 744]

; *ADS Aerospace Ltd v EMS Global Tracking Ltd* [2012] EWHC 2904 (TCC); and *PGF II SA v OMFS Co 1 Ltd* [2014] 1 W.L.R. 1386, [2013] EWCA Civ 1288.

(ii) Consequences of failing to mediate

19–009 [At note 49, delete reference to "CPR r.44.3(4)" and insert: page 745]

CPR Pt 44.2(4) (previously CPR Pt 44.3(4))

[Add to note 50: page 745]

As a general rule, silence in the face of an invitation to mediate will be unreasonable, regardless of whether a good reason for refusing to mediate existed: *PGF II SA v OMFS Co 1 Ltd* [2014] 1 W.L.R. 1386, [2013] EWCA Civ 1288.

[Add to note 51: page 745]

See also fn.40 above.

(iii) Mediation costs recoverable by the parties

[In the first sentence, substitute "Supreme Court Act 1981" with: page 746] **19–010**
Senior Courts Act 1981

4. TECHNOLOGY AND CONSTRUCTION COURT JUDGES (FORMERLY OFFICIAL REFEREES)

Practice.

[Amend fourth sentence of paragraph: page 747] **19–012**

The first CMC will be a directions hearing, for which purpose the parties must complete, exchange and return a case-management information sheet and either draft directions[71A] or a case-management directions form by no later than two days before the date on which the case-management conference is to take place.[72]

[71A] This is the preferred option: see the Guide, para.5.3.2.

[Add a further sentence after: "The first CMC ... the date on which the case-management conference is to take place": page 747]

For case-management conferences taking place after April 9, 2013, however, a new Pt 29.4, implemented following the Jackson Reforms, provides that the parties must endeavour to agree appropriate directions and must lodge agreed directions or their respective proposals to the court at least seven days before any case management conference.

5. COUNTY COURT JURISDICTION

[Amend third sentence of paragraph: page 749] **19–014**

However, proceedings (whether for damages or for a specific sum) may not be started in the High Court unless the value of the claim is £100,000 or more.[86]

[Add the following text to the end of this paragraph: page 749]

In addition, Akenhead J. recently gave guidance as to the circumstances in which it was appropriate to commence proceedings in the TCC in *West Country Renovations Ltd v McDowell*.[90A] Given the high volume of cases allocated to the TCC and the limited resources of the court, it was suggested that, generally, TCC claims valued at less than £250,000 (subject to certain limited exceptions) ought to be commenced in the county courts or other High Court centres outside London with TCC-designated judges.[90B]

[90A] [2013] 1 W.L.R. 416.
[90B] This is now reflected in the *Guide*, at para.1.3.1.

6. STATEMENTS OF CASE

(i) Statements of case

Composite or global claims.

19–021 [Delete the paragraph: "In *Bernhard's Rugby* ... applicable under the CPR." to sub-paragraph (c) and note 130 and replace with the following: page 755]

The authorities in relation to composite claims were recently reviewed by Akenhead J. in *Walter Lilly v Mackay*.[130] Drawing the authorities together, his Lordship set out the following principles:

(a) Ultimately, claims by contractors for delay or disruption related loss and expense must be proved as a matter of fact. Thus the contractor has to demonstrate on a balance of probabilities that first, events occurred which entitled it to loss and expense, secondly, that those events caused delay and/or disruption, and third, that such delay or disruption caused it to incur loss and/or expense.

(b) It is open to contractors to prove the three elements identified above with whatever evidence will satisfy the tribunal and the requisite standard of proof.

(c) There is nothing in principle wrong with a total or global cost claim, but there are added evidential difficulties in many such cases which need to be overcome. The contractor will generally have to establish that the loss which it has incurred would not have been incurred in any event, and that there are no other matters which contributed towards the loss claimed.

(d) The fact that one or a series of events or factors at the contractor's risk caused or contributed (or cannot be proved not to have caused or contributed) to the total or global loss does not necessarily mean that the claimant contractor can recover nothing, and whether or not that will be the case will depend on what the impact of those events or factors is.

(e) There is no need for the court to go down the global or total cost route if the actual cost attributable to individual loss causing events can be readily or practicably determined.

(f) It is wrong to say that a global award should not be allowed where a contractor has himself created the impossibility of disentanglement.

[130] [2012] EWHC 1773 (TCC), [473]–[492].

[Add note 131A to third paragraph: page 756]

On the other hand, if a global claim fails, ... individual losses to individual events.[131A]

[131A] See, for example, *Bluewater Energy Services BV v Mercon Steel Structures BV* [2014] EWHC 2132 (TCC), [1348]–[1367].

10. PREPARATION FOR TRIAL

(a) Preparation of evidence

Witness statements.

[Add the following text at the end of para.19–043: page 769]   **19–043**

The court may also give directions identifying or limiting the issues to which factual evidence may be directed, or limiting the length or format of witness statements.[185A]

[185A] See CPR Pt 32.2(3) which was implemented following the Jackson Reforms.

Experts' evidence.

[In the first paragraph of para.19–045, add the following text at the end of   **19–045** the sentence: "When applying for such permission ... and the identity of the expert (where practical)": page 770]

and, following the Jackson Reforms, an estimate of the costs of the proposed expert evidence and the issues which the expert evidence will address.

[In the first paragraph of para.19–045, add the following text at the end of the sentence: "The court has a discretion from any other party[196]": page 770]

and may also specify the issues which the expert evidence should address when granting permission.[196A]

[196A] CPR Pt 35.4(3).

19–046 [Add a new paragraph after: "The court may appoint an assessor ... that it attends court[203]": page 771]

The potential for experts to provide evidence concurrently, often known as "hot tubbing", is also now recognised as a result of the Jackson Reforms, following a successful pilot scheme in the TCC in Manchester. The procedure is in the discretion of the judge. It might therefore include the procedure currently set out in the TCC Guide.[203A] The procedure envisaged in the practice direction is for the judge to initiate discussion by asking the experts, in turn, for their views; for the parties' representatives to then have a chance to cross-examine; and then for the judge to summarise the experts' respective positions and to ask them to confirm or correct the summary given.[203B]

[203A] para.13.8.2(d).
[203B] CPR 35 PD11.4.

[Amend reference in note 208: page 772]

H.M. Malek, J. Auburn, R. Bagshaw, et al, *Phipson on Evidence*, 18th edn (London: Sweet & Maxwell, 2013)

Documents.

19–048 [Amend reference in note 219: page 774]

H.M. Malek, J. Auburn, R. Bagshaw, et al, *Phipson on Evidence*, 18th edn (London: Sweet & Maxwell, 2013).

(b) Disclosure and inspection of documents

19–049 [Add the following paragraphs after: "Any duty of disclosure ... not a party to the proceedings.[236]": page 775]

From April 1, 2013, the Jackson Reforms have also made significant changes to the CPR in respect of disclosure. Rule 31.5 has been replaced and the court will now consider a menu of possible disclosure options. For all multi-track claims, other than those which include a claim for personal injuries, unless the court orders otherwise, each party must file a disclosure report 14 days before the first CMC describing what documents exist or may exist that are or may be relevant, where and with whom those documents are or may be located, the broad range of costs that could be involved in giving standard disclosure, and which form of disclosure is thought to be appropriate.[236A] The parties are then obliged to seek to agree a proposal in relation to disclosure that meets the overriding objective not less than seven days before the first CMC.[236B] If agreement is reached, then the court may approve the parties' proposals without the need for a hearing.[236C]

Perhaps most importantly, CPR Pt 31.5(7) now provides a range of possible forms of disclosure which the court may adopt. These include: an

PREPARATION FOR TRIAL

order dispensing with disclosure; an order that a party disclose the documents on which it relies, and requests any specific disclosure it requires from any other party; an order that directs disclosure on an issue by issue basis where practicable; a "train of enquiry" order[236D]; an order for standard disclosure; or any other order that the court considers appropriate.

[236A] CPR Pt 31.5(3).
[236B] CPR Pt 31.5(5).
[236C] CPR Pt 31.5(6).
[236D] *Compagnie Financiere du Pacifique v Peruvian Guano Company* (1882) 11 QBD 55.

Legal professional privilege.

[Delete note 237 and replace with the following: page 775]   **19–050**

See *R (on the application of Prudential Plc) v Special Commissioner of Income Tax* [2013] 2 W.L.R. 325, where the Supreme Court (Lords Sumption and Clarke JJ.S.C dissenting) confirmed that legal advice privilege could not be relied upon in respect of communications with professionals other than lawyers (in that case, an accountant) even where they were competent to offer legal advice. See also, *Walter Lilly & Co Ltd v Mackay* [2012] EWHC 649 (TCC), in which Akenhead J. held that documents passing between the defendant and a claims consultant company did not attract legal professional or legal advice privilege.

(c) Preliminary point

[Add to note 253: page 778]   **19–052**

See also, *Barclays Bank Plc v Unicredit Bank AG* [2011] EWHC 3013 (Comm) where the court declined to order the trial of preliminary issues where although they would dispose of an important aspect of the case, it was very uncertain whether they would ultimately achieve the important and salutary objective of an overall saving of time and costs.

(e) View

Delay in proceedings.

[Add a new paragraph after: "In addition the courts ... a thing of the past.": **19–055** page 781]

The court's discretion to give relief from sanctions has been modified by the Jackson Reforms. The court is now to consider all the circumstances of the case, but specifically the need for litigation to be conducted efficiently and at proportionate cost and to enforce the compliance with rules, practice directions and orders.[271A] Following the decision in *Mitchell MP v News*

Group Newspapers Ltd[271B] it was thought that even a relatively minor failure to comply with orders or rules could result in a refusal to grant relief from sanctions under CPR Pt 3.9(1). However, in *Denton v TH White Ltd*[271C] the Court of Appeal clarified the operation of the rule. Thus an application for relief should consider three factors. The first being to identify and assess the seriousness and significance of the failure to comply with the relevant order or rule. The second factor is to consider why the default occurred and the third stage is to evaluate "all the circumstances of the case, so as to enable [the court] to deal justly with the application, including [factors one and two]." In particular the court held that if the breach is serious and significant and there is no good reason for the failure to comply with the relevant order or rule, the application for relief will not automatically fail. The court must go on to consider all the circumstances of the case, giving particular weight to the first two factors. Thus, for example, the promptness of the application for relief and past or current breaches of order or rules may well be relevant and taken into account as part of "all the circumstances of the case."

[271A] CPR Pt 3.9(1).
[271B] [2013] EWCA Civ 1537; [2014] B.L.R. 89.
[271C] [2014] EWCA Civ 906; [2014] B.L.R. 547.

(f) Settlement of actions

19–056 [Amend reference in note 272: page 781]

R. Turner, J.I. Winegarten, His Honour Judge M. Kershaw (eds), *Atkins Encyclopaedia of Court Forms*, 2nd edn (London: LexisNexis, 2014), Vol.12(1),

11. SET-OFF AND COUNTERCLAIM

Exclusion of right of set-off.

19–064 [Add the following sentence in note 302 after the sentence "By contrast ... the UCTA reasonableness test.": page 786]

A "no set-off" clause in a seller's standard terms and conditions was also held to satisfy the test of reasonableness in *FG Wilson (Engineering) Ltd v John Holt & Co (Liverpool) Ltd* [2012] EWHC 2477 (Comm), (a conclusion not affected by the decision in the appeal ([2014] 1 W.L.R. 2365)).

12. SUMMARY JUDGMENT AND INTERIM PAYMENT

Stay of execution.

19–070 [Amend reference in note 322: page 789]

Civil Procedure 2014

SUMMARY JUDGMENT AND INTERIM PAYMENT

Interim payment.

[Amend reference in note 330: page 790]   **19–071**

Civil Procedure 2014, Vol.2, para.15–127,

13. INTEREST

Interest ordinarily awarded.

[Amend reference in note 346: page 792]   **19–073**

Civil Procedure 2014

Interest on judgment debts.

[In note 357, delete the reference to "CPR Pt 44.3(6)(g)" and insert: page 793]   **19–076**

CPR Pt 44.2(6)(g) (previously CPR Pt 44.3(6)(g))

[In note 358, delete "CPR Pt 44.12(2)" and insert: page 793]

CPR Pt 44.9(4) (previously CPR Pt 44.12(2)

[Insert a further sentence after the sentence: "Where judgment is entered ... discussed above.[359]": page 794]

The rate of interest awarded for the period between the judgments on liability and quantum will fall to be considered as a matter for the court's discretion under s.35A of the Senior Courts Act 1981[359A] and will not be determined by the prevailing rate under s.17 of the Judgments Act 1838.[359B]

[359A] See para.19–072 above, and following.
[359B] *Persimmon Homes (South Coast) Ltd v Hall Aggregates (South Coast) Ltd* [2012] EWHC 2429 (TCC).

[Add new paragraph to text: page 794]

Where judgment is given for a sum expressed in a currency other than sterling,[359C] the rate of interest on the judgment debt is a matter for the court's discretion under s.44A of the Administration of Justice Act 1970. This discretion will be exercised so as to give effect to the compensatory principle and so the rate of interest awarded is likely to reflect prevailing commercial rates.[359D]

[359C] See *Miliangos v George Frank (Textiles) Ltd* [1976] A.C. 443.
[359D] *Novoship (UK) Ltd v Mikhaylyuk* [2014] EWCA Civ 908, per Longmore L.J., at [128]–[138]. The purpose of judgment debts under s.17 of the

CHAPTER NINETEEN—LITIGATION

Judgments Act 1838 is to compensate the claimant; that successive Lord Chancellors have failed to keep the sterling rate up-to-date has not altered that purpose. See also para 19–075 above.

14. COSTS

19–077 [Delete note 360: page 794]

[Delete the sentences from: "A pilot scheme has been set up" to end of paragraph and insert new paragraphs as following: page 794]

In addition, as a result of the Jackson Reforms there have been significant changes to the CPR in relation to costs which came into force on April 1, 2013. Indeed, CPR Pt 43 has been deleted in its entirety and replaced with New Costs Rules and Practice Directions, and Pts 44–48 have also been overhauled. Aside from the re-numbering of the various provisions, amongst the key changes are the following:

(a) A new regime is provided for the consequences of Pt 36 offers made on or after April 1, 2013.[361A] Where a claimant has made a Pt 36 offer which it subsequently betters at trial, in addition to the previous costs benefits the claimant will now also be awarded (unless the court considers it unjust to do so) an additional amount, not exceeding £75,000, which is 10 per cent of the sum awarded up to £500,000 and 5 per cent of the sum awarded above £500,000 up to £1 million. Where the claim is non-monetary, the percentage uplift will be applied to the sum awarded to the claimant in respect of costs.

(b) For cases commenced on or after April 1, 2013, it is clear that there will be a new emphasis on proportionality,[361B] which is likely to be particularly significant in construction cases. The new Pt 44.3(5) now includes a definition of proportionate costs and the new CPR Pt 44.3(2)(a) (previously CPR Pt 44.4) is amended to make clear that costs which are disproportionate in amount may be disallowed or reduced even if those costs were reasonably or necessarily incurred. Therefore, it is not enough that costs were reasonably or even necessarily incurred; they must also be "proportionate". Particularly where a dispute concerns, for example, a final account claim comprising a large number of modestly-valued variations or defects, this is likely to cause difficulties. This is particularly so given that, in accordance with the new rule, costs will only be proportionate if they bear a reasonable relationship to (amongst other things) the sums in issue in the proceedings and the complexity of the litigation.[361C] Thus Ramsey J. has said that "the new proportionality rule will dominate the way in which costs are assessed on a standard basis".[362D]

(c) For multi-track cases commenced on or after April 1, 2013, the addition of a new CPR Pt 3.12–3.18 and accompanying Practice Direction make provision for Costs Budgeting. The Rules do not apply to the Admiralty or Commercial Courts, or to Chancery, Mercantile and TCC cases with a value in excess of £2 million (excluding interest in costs), except where the court so orders. All parties other than litigants in person must (unless the court otherwise orders) file and exchange costs budgets for the litigation.[361E] The sanction for a failure to file a costs budget is draconian: the party in default will be assumed to have filed a budget comprising only the applicable court fees.[361F] The courts are prepared to apply this sanction strictly.[361G] The court will then likely make a costs management order setting out the extent to which the budgets are approved.[361H] Where such an order has been made, the successful party will only recover costs contained within its last approved budget unless there is good reason to depart from it.[361I]

(d) Important changes are also made in connection with conditional fee agreements. Where the claimant has entered into a CFA and taken out "after the event" insurance, the CFA uplift and ATE premium will no longer be recoverable from the losing defendant.[361J] Provision is made, however, for agreements based on payment to the party's lawyers of a certain percentage of the damages recovered.[361K]

[361A] Civil Procedure (Amendment) Rules 2013 (SI 2013/262) rr.14 and 22(7).
[361B] See Lord Neuberger of Abbotsford MR, Proportionate Costs: *http://www.judiciary.gov.uk/Resources/JCO/Documents/Speeches/proportionate-costs-fifteenth-lecture-30052012* [Accessed September 2013].
[361C] CPR Pt 44.3(5).
[361D] Ramsey J., Costs Management: A necessary part of the management of litigation: *http://www.judiciary.gov.uk/Resources/JCO/Documents/Speeches/costs-management-sixteenth-implementation-lecture-300512.pdf* [Accessed September 2013].
[361E] CPR Pt 3.13.
[361F] CPR Pt 3.14. *Mitchell v News Group Newspapers* [2013] EWCA Civ 1537; [2014] B.L.R. 89. However, this case has to be read in the light of *Denton v TH White Ltd* [2014] EWCA Civ 906; [2014] B.L.R. 547.
[361G] See *Mitchell v News Group Newspapers* [2014] 1 W.L.R. 795; [2013] EWCA Civ 1537 but which must now be read as subject to *Denton v TH White Ltd* [2014] EWCA Civ 906; [2014] B.L.R. 547. However, a failure to comply with the formal requirements of a costs budget will not necessarily render it a nullity: *Bank of Ireland v Philip Pank Partnership* [2014] 2 Costs L.R. 301; [2014] EWHC 284 (TCC) (failure to append a statement of truth).
[361H] CPR Pt 3.1.

Chapter Nineteen—Litigation

361I CPR Pt 3.18. See *Henry v News Group Newspapers Ltd* [2013] EWCA Civ 19 in which the Court of Appeal held that there was good reason to depart from the last approved budget under the provisions of a costs management pilot scheme but emphasised (at [28]) the difference between this and the new cost management scheme as from April 1, 2013. If the budget prescribed by a costs management scheme is likely to be exceeded, then formal approval ought to be sought from the court as soon as it was apparent that the original budget had been exceeded by more than a minimal amount, and it seems in any event prior to trial: *Elvanite Full Circle Ltd v AMEC Earth & Environmental* (UK) Ltd [2013] EWHC 1643 (TCC).

361J Legal Aid, Sentencing and Punishment of Offenders Act 2012 ss.44, 46.

361K Legal Aid, Sentencing and Punishment of Offenders Act 2012 s.45; Damages-Based Agreements Regulations 2013 (SI 2013/609).

Security for costs.

19–078 [Amend reference in note 367: page 795]

Civil Procedure 2014

[Amend reference in note 372: page 795]

Civil Procedure 2014

(a) Costs are discretionary

19–080 [In note 386, delete "CPR Pt 44.3(1)" and insert: page 797]

CPR Pt 44.2(1) (previously CPR Pt 44.3(1))

[In note 387, delete "CPR Pt 44.3(2)(a)" and insert: page 797]

CPR Pt 44.2(2)(a) (previously CPR Pt 44.3(2)(a))

[In note 388, delete "CPR Pt 44.3(4)" and insert: page 797]

CPR Pt 44.2(4) (previously CPR Pt 44.3(4))

[In note 390, delete "CPR Pt 44.4(1)(a)" and insert: page 797]

CPR Pt 44.3(1)(a) (previously CPR Pt 44.4(1)(a))

Pt 20 party.

19–082 [In the sentence: "Given the broad discretion ... under the CPR." delete the reference to "CPR Pt 44.3" and insert: page 798]

CPR Pt 44.2 (previously CPR Pt 44.3)

[In note 399, delete the reference to "CPR Pt 44.14 and CPR Pt 48.7" and insert: page 798]

CPR Pt 44.11 and CPR Pt 46.8 (previously CPR Pt 44.14 and CPR Pt 48.7)

[In note 401, delete the reference to "Section 13" and insert: page 798]

Section 9 (previously Section 13)

[In note 403, delete the reference to "CPR Pt 44.3" and insert: page 799]

CPR Pt 44.2 (previously CPR Pt 44.3)

(e) Written offers

[Add to note 419: page 801] **19–088**

"More advantageous" means better in money terms by any amount: Pt 36.14(1A). If the claimant succeeds in obtaining a more advantageous judgment than that contained in a defendant's Pt 36 offer by any margin (however small), the normal rules on costs will apply. There is no "near-miss" rule for Pt 36 offers and one cannot be introduced by means of Pt 44.2(4)(c): *Hammersmatch Properties (Welwyn) Ltd v Saint-Gobain Ceramics & Plastics Ltd* (2013) 149 Con. L.R. 147; [2013] EWHC 2227 (TCC).

[Add a new paragraph after the sentence: "In considering whether it would be unjust ... the matters set out at Pt 36.14(4).": page 801]

In addition, where the claimant has made a Pt 36 offer after April 1, 2013 and goes on to obtain judgment at least as advantageous as the offer, as set out above it will also be awarded (unless the court considers it unjust to do so) an additional amount, not exceeding £75,000, which is 10 per cent of the sum awarded up to £500,000 and 5 per cent of the sum awarded above £500,000 up to £1 million. Where the claim is non-monetary, the percentage uplift will be applied to the sum awarded to the claimant in respect of costs.

(f) Costs where there are cross-claims

[Add the following sentence after: "The nature of a special order ... some **19–090** small proportion".[432]: page 802]

Indeed, the Court of Appeal in *Connell v Mutch (t/a Southey Building Services)*,[432A] whilst declining to disturb the Order of a judge at first instance to award the claimant the costs of the claim and the defendant the costs of a successful but smaller counterclaim, held that the making of an Order based on a proportion of the successful party's costs was best practice.

[432A] [2012] EWCA Civ 1589.

CHAPTER NINETEEN—LITIGATION

[Delete the sentence: "It is not thought that any of cases cited are affected by the CPR.": page 802]

[In note 433, delete the reference to "CPR Pt 44.3" and insert: page 802].
CPR Pt 44.2 (previously CPR Pt 44.3)

Chapter Twenty

THE JCT STANDARD FORM OF BUILDING CONTRACT (2011 EDN)

INTRODUCTION

Articles 7, 8 and 9.

[Add after "para.18–105 and following." page 834]   20–019

For discussion, albeit in a particular context, of the inter-relationship between adjudication and other forms of dispute resolution.[27A]

[27A] *Price v Carter* [2010] 128 Con. L.R. 124.

CONDITIONS

SECTION 1 DEFINITIONS AND INTERPRETATION

Clause 1.3: *"... nothing contained in the Contract Bills or the CDP Documents shall override or modify the Agreement or these Conditions ..."*

[Add new note 38A: page 847]   20–029

.they continued to have no effect.[38A]

[38A] For a recent illustration of a court giving effect to Clause 1.3, so that Clauses 4.9 and 4.10 of the JCT Design and Build Contract took precedence over conflicting requirements in the Employer's Requirements, see *Fenice Investments Inc. v Jerram Falkus* [2010] 128 Con. L.R. 124.

Steps which limit the conclusiveness of the Final Certificate.

[Add new paragraphs: page 849]   20–036

In *University of Brighton v Dovehouse Interiors Ltd*.[49A] the Court gave guidance as to what was required by way of commencement of proceedings in order to stop the Final Certificate becoming conclusive.

It was held that adjudication proceedings for the purpose of the contract are commenced by the issue of an adjudication notice under para.1 of the Scheme. It was also held that the invalidity of the referral in that case did not mean that the notice of adjudication was invalid for the purpose of commencing proceedings under Clause 1.9.2.

[49A] [2014] B.L.R. 432.

SECTION 2 CARRYING OUT THE WORKS

Clause 2.4: "...*regularly and diligently proceed with...the same*..."

20–056 [Add new third paragraph: page 874]

In *Sabic UK v Punj Lloyd* [72A] the court (in relation to a different form of contract) held that the obligation of due diligence imported (but was not limited to) an obligation to carry out and complete the works industriously, assiduously, efficiently and expeditiously.

[72A] [2014] B.L.R. 43.

Clause 2.28.1.2: "*completion of the Works or of any Section is likely to be delayed thereby beyond the Completion Date*"

20–100 [Add new paragraph after: "It remains to be seen...followed by the English Courts": page 887]

The majority approach to assessing extensions of time (certainly under Clause 25) in *City Inn* reflects what has been accepted as the orthodox *Malmaison* approach up to the point at which the court apportions the delay. It is to be noted that none of the authorities cited by either the Outer or Inner House deal with apportionment of time (as opposed to money).[101A] Lord Carloway's approach is precisely the reverse: it rejects the *Malmaison* approach by appearing to suggest that it is not necessary to show that the relevant event is an actual cause of delay but is orthodox on the issue of apportionment.

The judgment of the Inner House in *City Inn* together with the proper approach to assessing delay pursuant to an extension of time clause (not however with similar wording to the JCT version) was considered in *Adyard Abu Dhabi v SD Marine Services*.[101B] Hamblen J. accepted that the *Malmaison* approach was correct and said in terms that Lord Carloway's approach does not reflect English law (at [286]).

Hamblen J. did not have to consider the position where (as a matter of fact) there was concurrent delay. However, in *Walter Lilly v Mackay*, the court also took a different approach to the Scottish courts in relation to an earlier version of the JCT Form.[101C] It held that:

(1) Clause 25.3.1 deals with extensions of time being granted prior to practical completion to the extent that completion of the works is

likely to be delayed by events which are Relevant Events under the contract. The extension is to be fair and reasonable having regard to any of the Relevant Events under Clause 25.3.3. This contract requires consideration of what critically delayed the works as they went along as opposed to a purely retrospective exercise (see paras 362–365).

(2) Where a period of delay has two effective causes, one of which entitles the contractor to an extension of time being a Relevant Event and one of which does not, the Contractor is entitled to a full extension of time. The Relevant Events would otherwise amount to acts of prevention and it would be wrong to construe Clause 25 on the basis that the Contractor should be denied a full extension of time in those circumstances. The fact that the Architect has to award a fair and reasonable extension does not imply that there should be some apportionment in the case of concurrent delays (see para.370).

[101A] For a criticism of the reasoning in *City Inn*, see "Apportionment and the common law: has *City Inn* got it wrong", B. McAdam (2009) 25 (2) Const. L.J. 79, 95.
[101B] [2011] EWHC 848 (Comm), [257–292].
[101C] [2010] B.L.R. 503.

Clause 2.30: "*When practical completion of the Works or a Section is achieved*"

[Add at the end of note 121: page 894] 20–120

The latter case was applied by Coulson J. in *Hall v Van Der Heiden* [2010] EWHC 586 (TCC).

[Add new paragraph after: "(d)minor items of work left incomplete, on "de minimis" principles.[125]"": page 894]

In *Walter Lilly v Mackay*, the court held that "Practical Completion" means completion for all practical purposes and what completion entails depends upon the nature, scope and contractual definition. Clause 17.1 (in an earlier version of the Form) requires the architect to certify when Practical Completion has been achieved. De minimis snagging is not a bar to practical completion (see para.372).[125A] This appears to support the approach set out in the text.

[125A] [2012] B.L.R. 503.

Chapter Twenty—The JCT Standard Form of Building Contract (2011 edn)

Clause 2.38: *"...at no cost to the Employer....deduction...from the Contract Sum."*

20-148 [Add new paragraph: page 904]

In *Oksana Mul v Hutton Construction Ltd* [158A] the court considered the meaning of the phrase "appropriate deduction" under Clause 2.30 of the Intermediate Form of Contract (equivalent to Clause 2.38). This meant a deduction which was reasonable in all the circumstances and could be calculated by reference to one or more of the following, amongst possibly other factors:

(a) the Contract rates/priced schedule of works/specification;

(b) the cost to the contractor of remedying the defect (including the sums to be paid to third party sub-contractors engaged by the contractor);

(c) the reasonable cost to the employer of engaging another contractor to remedy the defect; or

(d) the particular factual circumstances and/or expert evidence relating to each defect and/or the proposed remedial works

[158A] [2014] B.L.R. 529.

SECTION 4 PAYMENT

Clauses 4.12 and 4.13: Interim payments—final date and amount: Pay Less Notice

Clauses 4.12.5 and 4.1.3.1.

20-234 [Add at the end of note 208: page 951]

See also, *Balfour Beatty v Modus Corovest* [2008] All E.R. (D) 157.

Clauses 4.23–4.26

Loss and Expense. Nature of provisions for loss and expense.

20-273 [Add new paragraph after: "Clause 4.26 preserves any other rights and remedies which the Contractor may possess including, in particular, his right to claim damages": page 961]

In *Walter Lilly v Mackay*, the court gave general guidance as to Clause 26 of an earlier version of the JCT Form, the predecessor of these clauses.[230A]

(1) Clause 26.1 sets out two conditions precedents to any entitlement to recover loss and expense under Clause 26. The contractor had to make a timely application to the Architect and had to provide details of loss and expense to the quantity surveyor. It is difficult and undesirable to lay down any general rule as to what in every case needs to be provided. It is legitimate to consider what knowledge and information the Architect already had (see paras 464–470).

(2) A contractor can recover head office overheads and profit lost as a result of delay on a construction project caused by factors which entitle it to loss and expense. It is necessary for the contractor to prove on a balance of probabilities that if the delay had not occurred then it would have secured work or projects which would have produced a return over and above costs representing a profit and/or contribution to head office overheads. The use of a formula such as Emden or Hudson is a legitimate and helpful way of ascertaining, on a balance of probabilities, what that return can be calculated to be. The ascertainment process under Clause 26 does not mean that the assessor must be certain that the overheads and profit have been lost (see para.543).

(3) The contractor had to demonstrate that the regular progress of the works or a part thereof had been materially affected by any one or more of the matters referred in Clause 26.2 and that as a consequence its sub-contractor had been delayed and disrupted. Further, as a result it was put in a position in which it faced a substantial and broadly meritorious claim which it was reasonable to settle. It is open to the court in appropriate circumstances to make an apportionment of the settlement sum, if, and to the extent that, it can be confident that the sum allowed represents a realistic and reasonable allowance which can safely be attributed to the matters for which the defending party is liable (see paras 563–565).

However, it has been held that, although a provision such as Clause 4.23 prevents double recovery, it does not automatically exclude a claim under another clause if the conditions of that clause are satisfied.[230B]

[230A] [2012] B.L.R. 503.
[230B] *WW Gear v McGee Group* [2012] 1 B.L.R. 355.

SECTION 6 INJURY, DAMAGE AND INSURANCE

Clauses 6.1–6.3: Injury to Persons and Property

20–313 [Add at the end of note 263: page 985]

For a recent discussion of the *Canada Steamship* principle, see *Greenwich Millennium Village Ltd v Essex Services Group Plc.* [2014] 1 W.L.R. 3517, CA.

SECTION 8 TERMINATION

Clause 8.3.1: *"without prejudice to any other rights and remedies"*

20–362 [Add after: "It is clear that this express reservation of any other rights or remedies preserves the normal rights at common law of such Party against the other.[298]": page 1012]

It has been held that a provision of this kind does not put a party to its election between operating the termination machinery and claiming damages for breach of contract.[298A]

[298A] *Aviva Insurance Ltd v Hackney Empire* [2013] B.L.R. 57.

Clause 8.5: Insolvency of Contractor

20–374 [Add at the end of the paragraph: page 1015]

However, although provisions such as Clauses 8.5.3.1 and 8.7.3 (as to which see below) have important effects on payment, it has been held that these clauses do not have the effect of prevailing over the paying party's obligation to comply with an Adjudicator's decision.[313A]

[313A] See *Straw Realisations (No.1) Ltd v Shaftesbury House Developments Ltd* [2010] All E.R. (d) 196.

SECTION 9 SETTLEMENT OF DISPUTES

Clause 9.2: "...*arises under this Contract*..."

20–405 [Delete the words ""for example, dispute as to ....Clause 9.2" and substitute: page 1027]

...for example, disputes as to the initial existence of the Contract are probably excluded from the ambit of art.7 and Clause 9.2 and it has been held that a claim under the Misrepresentation Act 1967 arose under that Act, not 'under this Contract' as required by art.7.[335]

[Delete note 335 and substitute: page 1027]

[335] *Hillcrest Homes Limited v Beresford and Curbishley Ltd* [2014] Con. L.R. 179.

CHAPTER TWENTY-ONE

THE INFRASTRUCTURE CONDITIONS OF CONTRACT

[Insert new note 0A to the end of the first sentence of the second paragraph of 21–001: page 1074] **21–001**

It is not yet known what effect this will have on the use of the Conditions and the extent to which they will be revised under their new ownership.[0A]

[0A] Since the First Supplement to the 9th edn, of July 2013, the Restructuring Group (established by the Users Forum of the sponsoring bodies, the ACE and CECEA) has produced a new and substantially revised version of the form, so drafted as to be suitable for use with a wider range of construction activities either within the UK or abroad. The new draft has been approved by its sponsors and was published in October 2014. It is intended that the new draft will be used as a template for updating other forms within the existing suite of ICC documents, all of which (with the exception of the "blue form" of sub-contract) have been inherited from the ICE as noted in the main work. It is likely that the next documents to be produced will be a sub-contract form and a target cost version of the main contract form. The notes below outline the principal features of the new form.

Notes on new ICC conditions

The Conditions are substantially shorter, being about half the previous length. The number of clauses is also reduced to 20 together with four supplementary clauses. The objective was to maintain the existing balance of risk and the "feel" of the previous versions, where appropriate by use of the same or similar wording. Some of the features of the earlier form have been maintained, such as definitions and interpretation in Clause 1, whilst the 70 clauses of the earlier version have been re-arranged and collated into fewer but broader clauses containing related provisions previously distributed throughout the form. Provisions dealing with insurance and termination are substantially repeated with some editing; whilst provisions dealing with risk, valuation and payment have been substantially modified. The very extensive provisions in the earlier editions dealing with resolution of disputes have been greatly simplified and re-written.

The many provisions under the earlier form dealing with risk have been collected together in one clause dealing both with risk in regard to the physical works and in regard to performance. Thus many provisions under the existing form giving rise to additional payment are collected together under the heading "Employer's Risks"; whilst those giving rise to an allowance of time only without additional payment are similarly collected under the heading "Shared Risks". For the measurement and valuation of the works, a decision was taken that the traditional practice of treating all billed work as subject to re-measurement was no longer justifiable. Instead the default position is that billed items are to be treated as lump sums with the option of providing for re-measurement or, via a supplementary clause, using milestone sums to be payable only upon achievement of stated criteria. The contract is thus intended to achieve both greater flexibility and certainty of outcome. For changes to the work, while the traditional mode of valuation using rates and prices contained in the bill is maintained, alternative provisions for advance valuation by the contractor are provided including any consequential extensions of time.

The now traditional and very lengthy provisions governing responsibility for nominated sub-contractors have been entirely removed; however, unlike the JCT form, from which nomination has entirely disappeared, provision for nomination is maintained on the basis that the contractor must accept full responsibility subject to reasonable objection. In the event of such objection and inability to proceed with the intended nominee, new provisions entitle the employer to engage the intended sub-contractor as a direct contractor, thus avoiding the possibility of substantial delay where procurement rules would otherwise require a lengthy re-tendering process.

Provisions dealing with programme, delay and extension of time substantially follow the earlier edition but with re-arrangement. With regard to claims for additional payment, all previous editions have made provision either for additional cost claims or for the engineer to adjust the rates applicable to items of work affected by variations. Both of these are now superseded by a single provision allowing the Engineer to determine additional cost incurred through delay or disruption of the works.

Where disputes arise it is recognised that, in a UK jurisdiction and increasingly in foreign jurisdictions, there will be a statutory right to adjudication leading to an enforceable but reviewable decision. In the UK and most other jurisdictions the adjudicator is allowed 28 days to give a decision, which may itself be extended. Given the longstanding tradition of an initial decision being rendered by the Engineer, the drafting team decided to re-introduce this as an option which might be invoked by either party at any time, with the engineer being required to give a decision on a matter referred to him within 14 days. Like an adjudication decision, the engineer's decision is to be temporarily binding pending the decision of an adjudicator or an Arbitral Tribunal; however, if not so challenged, the decision will remain binding and may be enforced. Alternative provisions allow for adjudication, conciliation or arbitration generally at the option of the referring party.

The form also contains new measures which have been found to be beneficial in other forms of contract. Thus, there is provision for collaboration and early warning designed to lead pro-actively to the adoption of measures aimed at avoiding or mitigating delay or additional cost. This may lead either to agreement or to the issue of appropriate instructions by the engineer. Another new measure, by a supplementary clause, is provision for Employer Furnished Materials to be incorporated into the works by the contractor. The form is published together with a Form of Tender and Appendix, a Form of Agreement, a Form of Bond and optional Index-based fluctuation clauses.

Chapter Twenty-Two

ENGINEERING AND CONSTRUCTION CONTRACT (NEC 3) AND INTERNATIONAL FEDERATION OF CONSULTING ENGINEERS CONDITIONS (FIDIC)

1. ENGINEERING AND CONSTRUCTION CONTRACT (NEC3)

[Delete the paragraph "It is clearly intended...be most welcome by users of the form" and substitute: page 1254] **22–007**

A contractor can rely upon the failure of the project manager to notify the event under Clause 61.1, even if the project manager reasonably believed it was not a compensatory event.[23] This decision, if correct, deprives the condition precedent of much of its force and effect. It is submitted that the purpose of the clause is to enable the parties to deal expeditiously with claims when the relevant facts can be readily ascertained. This decision means that the contractor can pursue compensation events long after their occurrence, even where the project manager had no reason to think the event in question qualified for compensation.

[23] *Northern Ireland Housing Executive v Healthy Buildings Ltd* [2014] NICA 27; 153 Con. L.R. 87.

2. INTERNATIONAL FEDERATION OF CONSULTING ENGINEERS (FIDIC)

[Add note 30A to the end of the sentence "Clauses 15 and 16 contain general provisions allowing termination by employer and contractor respectively.": page 1260] **22–013**

[30A] Provided the notice served under Clause 15.2 is received by the contractor and it is clear and unambiguous that it is being served pursuant to that clause, it will be effective. Thus such a notice may be effective if sent to the contractor's site office, rather than to its head or main office. Compliance with Clause 1.3 (which stipulates the address for service of, amongst other matters, notices on the contractor) is not an indispensable requirement of Clauses 1.3 or 15.2. See *Obrascon Huarte Lain SA v Her Majesty's Attorney General* [2014] EWHC 1028 (TCC); [2014] B.L.R. 484.

22–014 [Add to note 32, before the reference to *W.W. Gear Construction*: page 1261]

In *Obrascon Huarte Lain SA v Her Majesty's Attorney General* [2014] EWHC 1028 (TCC); [2014] B.L.R. 484 it was held that Clause 20.1 creates a condition precedent but that the requirement that the contractor had to be aware of the event or circumstance giving rise to the claim or should have been so aware should be construed broadly given its serious effect on what could otherwise be good claims for breach of contract by the Employer, by way of example.

INDEX

This index has been prepared using Sweet and Maxwell's Legal Taxonomy. Main index entries conform to keywords provided by the Legal Taxonomy except where references to specific documents or non-standard terms (denoted by quotation marks) have been included. These keywords provide a means of identifying similar concepts in other Sweet and Maxwell publications and on-line services to which keywords from the Legal Taxonomy have been applied. Readers may find some differences between terms used in the text and those which appear in the index. Suggestions to *sweetandmaxwell.taxonomy@thomson.com*. **(All references are to paragraph number).**

Acceptance
    'battle of forms', 2–017
    certainty of terms, 2–021
    conduct, by, 2–027
    contract to negotiate, 2–024
    lengthy negotiations, 2–018
    non-completion, 4–012
    post, telex, fax or -mail, by, 2–028
    price, 2–023
    'subject to contract', 2–020
Access (construction sites)
    subcontractors, 13–057
ACE conditions of engagement
    generally, 1–024
Addition of parties
    limitation periods, 16–022
Adjudication
    adjudicators
        appointment, 18–021
        fees and expenses, 18–029
        jurisdiction, 18–022
    approach of courts, 18–018
    breach of human rights, 18–051
    breach of natural justice, 18–050
    construction contracts
        excluded, 18–012
        meaning, 18–009
        PFI, 18–012
    enforcement
        implied obligation to comply, 18–032
        stay of execution, 18–034
        summary judgment, 18–033
    excluded contracts, 18–012
    grounds for opposition to enforcement
        breach of human rights, 18–051

        breach of natural justice, 18–050
        mistaken answer to an issue, 18–048
        no dispute, 18–042—18–045
        no jurisdiction over an issue, 18–046
    human rights, 18–051
    implied obligation to comply, 18–032
    introduction, 3–040
    jurisdiction, 18–022
    limitation periods, 16–024
    mistaken answer to an issue, 18–048
    natural justice, 18–050
    no dispute
        disputes 'under the contract', 18–043
        existence of dispute, 18–045
        introduction, 18–042
    no jurisdiction over an issue, 18–046
    procedure
        appointment of adjudicator, 18–021
        fees and expenses, 18–029
        jurisdiction, 18–022
        timing of referral, 18–020
    Standard Building Contract with
        Quantities, 20–405
    stay of execution, 18–034
    summary judgments, 18–033
    timing of referral, 18–020
Adjustment
    Standard Building Contract with
        Quantities
        completion, 20–100
Agents
    architects
        misconduct, 14–026
    corporations, 2–042

Index

Aliens
 capacity, 2–039
Alternative dispute resolution
 See also **Arbitration**
 generally, 17–001
 introduction, 1–038
Anti-social behaviour
 high hedges
  complaints procedure, 16–073
  definitions, 16–072
Appeals
 arbitration
  generally, 17–064—17–065
  leave to appeal, 17–066
Arbitration
 appeals to High Court
  generally, 17–064—17–065
  leave to appeal, 17–066
 appointment of arbitrators
  generally, 17–007
  judges, and, 17–020
 arbitration agreements
  definition, 17–004
  nature of contractual procedure, 17–007
  validity, 17–030
 arbitrators
  appointment, 17–007
  architect as, 17–022
  disqualification for bias, 17–021—
  judges as, 17–020
  jurisdiction, 17–010—17–013
  procedural powers, 17–039
 control by court
  appeals to High Court, 17–064
  extensions of time, 17–053
  injunction to restrain proceedings, 17–056
  interim injunctions, 17–061
  setting aside award, 17–057
 extensions of time, 17–053
 injunction to restrain proceedings, 17–056
 interim awards, 17–043
 interim injunctions, 17–061
 international arbitration (ICC Rules)
  proper law, 17–075
 introduction, 17–001
 jurisdiction
  challenge to, 17–017—17–018
  fraud, 17–016
  generally, 17–010—17–013
 legislative background, 17–001
 obligations of parties, 17–039
 peremptory orders, 17–040
 procedure
  arbitrators powers, 17–039
  interim awards, 17–043
  obligations of parties, 17–039
  peremptory orders, 17–040
 rectification, 17–013
 serious irregularity, 17–057

 setting aside award, 17–057
 stay of proceedings
  no dispute, 17–031
  valid arbitration agreement, 17–030
  'step in the proceedings', 17–029
Arbitration agreements
 definition, 17–004
 nature of contractual procedure, 17–007
 validity, 17–030
Arbitration clauses
 unfair contract terms, and, 17–013
Arbitrators
 appointment
  generally, 17–007
  judges, and, 17–020
 disqualification for bias, 17–021—
 judges as, 17–020
 jurisdiction, 17–010—17–013
 procedural powers, 17–039
Architects
 See also **Architects (certificates)**
 advice on contract, 14–044
 agents, as
  misconduct, 14–026
 architects duties
  advice on contract, 14–044
  breach of duties to employer, 14–055
  duration, 14–055
  estimates, 14–039
  knowledge of law and practice, 14–037
 authority as agent
  exceeding, 14–020—14–023
  introduction, 14–013
  supervision, 14–016
  tenders, 14–014
  variation, 14–017
 bankruptcy, 14–065
 breach of duties to employer
  introduction, 14–055
  limitation periods, 14–059—14–061
  negligent survey, 14–057
  tortious liability, 14–058
  unauthorised subcontracting, 14–056
 bribes, 14–026
 care and skill
  degree required, 14–029
  expert evidence, 14–032
  extent of professional knowledge, 14–030
  introduction, 14–028
  state of the art, 14–030
  warranty of fitness, 14–031
 completion of works, 14–062
 conflict of duty, 14–027
 conflict of interest, 14–027
 construction materials, 14–043
 contract with employer
  duty to act fairly, 14–012
  introduction, 14–010
  limited companies, 14–011
 contractual liability, 14–024

copyright in plans and design
  introduction, 14–078
  licence to reproduce, 14–079—14–080
  measure of damages, 14–082
  remedies fro breach, 14–081
death, 14–064
defective premises, 14–087
delegation, 14–033—14–034
design
  copyright, 14–078—14–082
  duties, 14–041
  drawings, 14–040
duration of duties
  bankruptcy, 14–065
  completion of works, 14–062
  death, 14–064
  resignation, 14–063
  termination, 14–063
duty to act fairly, 14–012
EEA state nationals, 14–006
estimates, 14–039
examination of site, 14–038
exceeding authority
  architect's position, 14–022
  damages, 14–023
  employer's position, 14–020
knowledge of law and practice, 14–037
limitation periods
  contractual claims, 14–060
misconduct
  bribes, 14–026
  secret commission, 14–026
professional conduct, 14–008
registration
  professional conduct, 14–008
  removal for non-payment of fees, 14–007
  retention of name on register, 14–007
retention of name on register, 14–007
secret commission, 14–026

Architects (certificates)
binding and conclusive
  expiration of time, 5–028
  whether and to what extent, 5–024
challenges to
  mistake, 5–039
conditions precedent
  introduction, 5–014
expiration of time, 5–028
interim certificates
  Standard Building Contract with Quantities, 20–234
mistakes, 5–039
not properly made
  mistake, 5–039
payment without
  prevention by employer, 5–020
retention money, 5–013
Standard Building Contract with Quantities
  interim certificates, 20–234

Assignment
benefit of contract, of
  contractors, by, 13–005
  introduction, 13–001
burden of contract
  contractors, by, 13–004
  introduction, 13–001
causes of action, 13–020
champerty, 13–020
contractors, by
  benefit, of, 13–005
  burden, of, 13–004
contractors of benefit, by
  equity, in, 13–007
  fraud, 13–013
  introduction, 13–005
  s 136 LPA 1925, under, 13–006
contractors of burden, by
  vicarious performance, 13–004
employers of benefit, by
  causes of action, 13–020
  invalid assignment, 13–019
  maintenance and champerty, 13–020
  third party rights, 13–022
  warranties, 13–021
equitable assignment, 13–007
fraud, 13–013
introduction, 13–001
maintenance and champerty, 13–020
statutory assignment, 13–006
third party rights, 13–022
vicarious performance, 13–004
warranties, 13–021
"Battle of forms"
formation of contract, 2–017
Bills of quantities
extra work, 4–027
payment clauses, 4–027
Bonds
conditional bonds, 11–035
on-demand bonds, 11–036
Breach of contract
causation
  'but for' test, 9–062
  claimant and defendant, 9–070
  concurrent causes, 9–059
  contractual claims, 9–058
  relevance of common law principles, 9–060
concurrent causes
  claimant and defendant, 9–070
  generally, 9–059
damages
  date of assessment, 9–009
  Hadley v Baxendale rule, 9–006
  Hadley v Baxendale rule
    second limb, 9–005—9–006
  introduction, 9–001
  liquidated damages
    See also **Liquidated damages**

Index

defences to claims, 10–005—10–020
Bribery
    architect, 14–026
Building regulations
    contravention, 16–028
    drains, 16–025
    generally, 16–025
    legislative background, 16–025
    local authority's liability, 16–031
    penalties, 16–028
    sewers and drains, 16–025
"'But for' test"
    causation, and, 9–062
Capacity
    aliens, 2–039
    corporations
        agents, 2–042
        generally, 2–042
    unincorporated associations, 2–047
Case management
    litigation, 19–001
Causation
    'but for' test, 9–062
    claimant and defendant, 9–070
    concurrent causes
        claimant and defendant, 9–070
        generally, 9–059
    contractual claims, 9–058
    relevance of common law principles, 9–060
Causes of action
    assignment of benefit by employers, 13–020
Certainty
    formation of contract, 2–021
Champerty
    assignment of benefit by employers, 13–020
Civil evidence
    witness statements, 19–043
Civil Procedure Rules
    litigation, 19–001
Clerical errors
    interpretation of contracts, 3–028
Collateral warranties
    extrinsic evidence, 3–018
Competitive dialogue procedure
    public procurement, 15–014
Completion
    payment clauses
        acceptance of work done, 4–012
        unpaid instalments, 4–014
Completion date
    express provision
        fixed date or period, 8–003
        nature of obligation, 8–004
    time of the essence, and
        notice making, 8–009
Concealment
    limitation periods, 16–019
Conciliation
    generally, 17–001

Concurrent causes
    claimant and defendant, 9–070
    extensions of time for delay, and
        apportionment, 8–027
    delay events, 8–026
    generally, 9–059
Conditional bonds
    generally, 11–035
Conditional payments
    introduction, 3–040
Conditions precedent
    architect's certificates
        introduction, 5–014
    extensions of time for delay, and
        generally, 8–030
Conduct
    formation of contract, 2–027
Consideration
    formation of contract, 2–001
Construction contracts
    contractual arrangements
        introduction, 1–022—1–024
        project management, 1–022
    design and build contracts
        standard forms, 1–028
        suitability for purpose, 1–030
    dispute resolution
        ADR, 1–038
    formation
        elements, 2–001
    Housing Grants Construction and Regeneration Act 1996
        excluded contracts, 18–012
        meaning, 18–009
        PFI, 18–012
    types, 1–022—1–024
Construction materials
    retention of title, 11–014
    vesting clauses, 11–015
Contra proferentem
    interpretation of contracts, 3–032
"Contracting authorities"
    public procurement, 15–009
Contractors
    assignment of benefit of contract
        equity, in, 13–007
        fraud, 13–013
        introduction, 13–005
        s 136 LPA 1925, under, 13–006
    assignment of burden of contract
        vicarious performance, 13–004
    claims for damages for breach of contract, for
        wasted expenditure, 9–038
        work partly carried out, 9–039
    claims for financial recovery under terms of contract
        'Hudson formula', 9–034
        loss of profit, 9–033

disclosure, 19–049
"Dynamic purchasing systems"
　public procurement, 15–025
Economic loss
　limitation periods, 16–013
Electronic auctions
　public procurement, 15–025
Electronic mail
　acceptance, 2–028
Emergencies
　extra work, 4–042
Employers
　assignment of benefit of contract
　　causes of action, 13–020
　　invalid assignment, 13–019
　　maintenance and champerty, 13–020
　　third party rights, 13–022
　　warranties, 13–021
　claims for damages for breach of contract
　　detective work, 9–047
　　discomfort, 9–053
　　distress, 9–053
　　inconvenience, 9–053
　　lack of inspection by architect, 9–055
　　offer to complete, 9–046
Enforcement
　adjudication
　　implied obligation to comply, 18–032
　　stay of execution, 18–034
　　summary judgment, 18–033
Engineering and Construction Contract
　compensation events, 22–007
　introduction, 1–022
Entire agreement clauses
　formation of contract, 2–032
Entire contracts
　generally, 4–003
　meaning, 4–003
　recovery of money paid, 4–007
　retention money, 4–005
Equal treatment
　public procurement, 15–028
Equipment
　retention of title, 11–014
　vesting clauses, 11–015
Equitable assignment
　benefit of contract by contractors, 13–007
Equitable remedies
　estoppel
　　convention, by, 12–003
　　introduction, 12–001
　　promissory, 12–004
　　representation, 12–002
　　types, 12–001
　injunctions
　　disputed forfeiture, and, 12–021
　　generally, 12–020
　rectification
　　common mistake, 12–012
　　expert determination, 12–015A

　　unilateral mistake, 12–013
　rescission
　　generally, 12–009
　specific performance
　　impossibility, 12–017A
　　introduction, 12–017
　variation, 12–009
　waiver
　　generally, 12–005
Estimates
　architect's powers and duties, 14–039
Estoppel
　convention, by, 12–003
　formation of contract, 2–031
　introduction, 12–001
　promissory, 12–004
　representation, 12–002
　types, 12–001
Exclusion clauses
　contractual liability, 3–078
　injunctions, 3–073A
　interpretation
　　introduction, 3–065
　legislative provisions
　　contractual liability, 3–078
　　introduction, 3–074
　　UCTA 1977, 3–075
　misrepresentation
　　fair and reasonable, 6–025
　　generally, 6–024
　negligence
　　introduction, 3–065
　　third parties, 3–073
Expert determination
　rectification, 12–015A
Expert witnesses
　reports, 19–045—19–046
Extensions of time
　apportionment, 8–027
　arbitration, 17–053
　concurrent causes of delay
　　apportionment, 8–027
　　delay events, 8–026
　conditions precedent
　　generally, 8–030
　retrospective extensions
　　approaches, 8–036
Extra work
　architect's final certificate, 4–052
　basis of claim
　　emergencies, 4–042
　bills of quantities contract
　　generally, 4–027
　dispute resolution
　　architect's final certificate, 4–052
　emergencies, 4–042
　final certificates, 4–052
　indispensably necessary works, 4–025
　lump sum contracts–exactly defined
　　bills of quantities contract, 4–027

lump sum contracts–widely defined
    indispensably necessary works, 4–025
measurement and value contracts
    generally, 4–030
    omissions, 4–033
rate of payment
    generally, 4–053
Extrinsic evidence
    blanks, 3–004
    collateral warranties, 3–018
    custom and usage, 3–013
    deletions from printed documents, 3–007
    introduction, 3–002
    surrounding circumstances, 3–011
Fax
    acceptance, 2–028
FIDIC conditions of contract
    introduction, 1–024
    termination, 22–013—22–014
Final certificates
    See also **Architects (certificates)**
    extra work, 4–052
    Standard Building Contract with
        Quantities
        conclusive effect, 20–036
Fitness for purpose
    completed works, 3–054
    exclusion of warranty, 3–053
    generally, 3–052
    sale of buildings, 3–058
Foreign currencies
    damages for breach of contract, 9–021
Forfeiture clauses
    effect
        relief, 11–011
    employer's default, 11–006
    insolvency of contractors, 16–041
    relief, 11–011
Formalities
    guarantees, 2–035
    suretyship, 2–035
Formation of contract
    acceptance
        conduct, by, 2–027
        post, telex, fax or -mail, by, 2–028
    aliens, 2–039
    'battle of forms', 2–017
    capacity of parties
        aliens, 2–039
        corporations, 2–042
        unincorporated associations, 2–047
    certainty of terms, 2–021
    conduct, 2–027
    consideration, 2–001
    contract to negotiate, 2–024
    corporations
        agents, 2–042
        generally, 2–042
    count-offers, 2–012
    elements, 2–001

entire agreement, 2–032
estoppel, 2–031
formalities
    guarantees, 2–035
    suretyship, 2–035
fraud, 2–001
guarantees, 2–035
illegality, 2–001
letter of intent, 2–008
misrepresentations, 2–001
mutual mistake, 2–001
notice of terms
    entire agreement, 2–032
    estoppel, 2–031
offer
    count-offer, 2–012
    letter of intent, 2–008
    rejection, 2–011
    tender, 2–004
price
    generally, 2–023
    'subject to contract', 2–020
suretyship, 2–035
tenders
    costs, 2–004
    generally, 2–003
unconditional acceptance
    'battle of forms', 2–017
    certainty of terms, 2–021
    contract to negotiate, 2–024
    lengthy negotiations, 2–018
    price, 2–023
    'subject to contract', 2–020
unincorporated associations, 2–047
Fraud
    See also **Fraudulent misrepresentation**
    arbitration jurisdiction, 17–016
    assignment of benefit by contractors,
        13–013
    formation of contract, 2–001
    limitation periods, 16–019
Fraudulent misrepresentation
    remedies, 6–012
Frustration
    recovery of sums, 6–047
GC/Works
    generally, 1–023
Guarantees
    formalities of contract, 2–035
    introduction, 11–025
    liability of surety, 11–026
    material alteration in contract, 11–033
    non-disclosure, 11–031
High hedges
    complaints procedure, 16–073
    definitions, 16–072
Houses
    duty to build properly
        'dwelling', 16–002
        generally, 16–002

measure of damages, 16–007
persons owing duty, 16–004
Housing Grants Construction and Regeneration Act 1996
adjudication
See also **Adjudication**
approach of courts, 18–018
introduction, 3–040
procedure, 18–020—18–029
construction contracts
excluded, 18–012
meaning, 18–009
PFI, 18–012
excluded contracts, 18–012
introduction
adjudication, 3–040
construction contracts, 3–040
construction operations, 3–040
generally, 3–040
payments, 3–040
payment
introduction, 3–040
withholding payment, 18–060
Scheme for Construction Contracts
introduction, 3–040
"Hudson formula"
damages for breach of contract, 9–034
Human rights
adjudication, 18–051
ICE conditions of contract
introduction, 21–001
overview
design and build contracts, 1–028
generally, 1–022—1–024
Illegality
formation of contract, 2–001
generally, 6–048
return of goods, 6–051
Implied contracts
public procurement, 15–045
Implied terms
co-operation, 3–046
dates for payment, 3–040
fitness for purpose
generally, 3–052
good quality, 3–052
Housing Grants, Construction and Regeneration Act 1996
dates for payment, 3–040
introduction, 3–040
notice of amount of payment, 3–040
prohibiting conditional payments, 3–040
referral of disputes, 3–040
stage payments, 3–040
suspension for non-payment, 3–040
withholding payment, 3–040
make contract work, 3–043
necessary implication
co-operation, 3–046
fitness for purpose, 3–052

good quality, 3–052
introduction, 3–042
make contract work, 3–043
payment clauses
conditional payments, 3–040
dates for payment, 3–040
introduction, 3–040
notice of amount, 3–040
stage payments, 3–040
suspension for non-payment, 3–040
withholding payment, 3–040
stage payments, 3–040
Supply of Goods and Services Act 1982, 3–041
suspension for non-payment, 3–040
withholding payment, 3–040
Impossibility
specific performance, 12–017A
Inconvenience
damages for breach of contract, 9–053
Indemnity clauses
interpretation, 3–065
limitation periods, 3–070
negligence
introduction, 3–065
Injunctions
arbitration, 17–056
exclusion clauses, 3–073A
forfeiture clauses
generally, 12–021
generally, 12–020
restraint of proceedings, in, 17–056
Innocent misrepresentation
damages, 6–020
generally, 6–007
rescission, 6–007
Insolvency
contractors
forfeiture clauses, 16–041
liens, 16–041
mutual dealings, 16–040
seizure, 16–041
Standard Building Contract with Quantities, 20–374
generally, 16–032
meaning, 16–032
Inspection
litigation, 19–049
Instalments
non-completion, 4–014
Instructions
negligence, 7–009
Insurance
Standard Building Contract with Quantities
contractors' liability, 20–313
Interest
arbitration awards, 17–043
judgment debts, 19–076
rule of practice, 19–073

Index

Interim certificates
  Standard Building Contract with
    Quantities
    payment, 20–234
Interim injunctions
  arbitration, 17–061
Interim orders
  public procurement, 15–036
Interim payments
  introduction, 3–040
  litigation, 19–071
International arbitration
  proper law, 17–075
Interpretation (contracts)
  blanks, 3–004
  collateral warranties, 3–018
  contra proferentem, 3–032
  correction of clerical error, 3–028
  correction of mistakes by construction, 3–024
  custom and usage, 3–013
  co-operation, 3–046
  deeds, 3–061
  exclusion clauses
    introduction, 3–065
  expressed intention
    extrinsic evidence, 3–002
    introduction, 3–002
  extrinsic evidence
    blanks, 3–004
    collateral warranties, 3–018
    custom and usage, 3–013
    deletions from printed documents, 3–007
    introduction, 3–002
    surrounding circumstances, 3–011
  fitness for purpose
    completed works, 3–054
    exclusion of warranty, 3–053
    generally, 3–052
    sale of buildings, 3–058
  good quality, 3–052
  implied terms
    Housing Grants, Construction and
      Regeneration Act 1996, 3–040
    necessary implication, 3–042
    statutory implication, 3–040
    Supply of Goods and Services Act 1982, 3–041
  implied terms (HGCRA 1996)
    dates for payment, 3–040
    introduction, 3–040
    notice of amount of payment, 3–040
    prohibiting conditional payments, 3–040
    referral of disputes, 3–040
    stage payments, 3–040
    suspension for non-payment, 3–040
    withholding payment, 3–040
  implied terms (SOGASA 1982), 3–041
  indemnity clauses, 3–065
  introduction, 3–001

  irreconcilable clauses, 3–034
  literal interpretation, 3–023
  make contract work, 3–043
  necessary implication
    co-operation, 3–046
    fitness for purpose, 3–052
    good quality, 3–052
    introduction, 3–042
    make contract work, 3–043
    ordinary meaning, 3–022
  'pay-when-paid' clauses, 3–040
  public authority contracts, 3–037
  reasonable meaning, 3–023
  recitals, 3–033
  risk clauses, 3–065
  rules
    contra proferentem, 3–032
    correction of clerical error, 3–028
    correction of mistakes by construction, 3–024
    irreconcilable clauses, 3–034
    ordinary meaning, 3–022
    reasonable meaning, 3–023
    recitals, 3–033
  statutory implication
    dates for payment, 3–040
    introduction, 3–040
    make contract work, 3–042
    notice of amount of payment, 3–040
    prohibiting conditional payments, 3–040
    referral of disputes, 3–040
    stage payments, 3–040
    suspension for non-payment, 3–040
    withholding payment, 3–040
    surrounding circumstances, 3–011
  usual terms–contractors
    fitness for purpose, 3–052
    good quality, 3–052
  usual terms–employer
    co-operation, 3–046
Judicial review
  public procurement, 15–042
Jurisdiction
  adjudication, 18–022
  arbitration
    challenge to, 17–017—17–018
    fraud, 17–016
    generally, 17–010—17–013
Laches
  liability of surety, 11–026
  material alteration in contract, 11–033
  non-disclosure, 11–031
Latent damage
  limitation periods, 16–017
Legal professional privilege
  disclosure, 19–050
Letters of intent
  formation of contract, 2–008
Liens
  insolvency of contractors, 16–041

Limitation clauses
See also **Exclusion clauses**
contractual liability, 3–078
injunctions, 3–073A
interpretation, 3–065
legislative provisions
  contractual liability, 3–078
  introduction, 3–074
  UCTA 1977, 3–075
negligence
  loss caused, 3–065
  third parties, 3–073
Limitation periods
addition of parties, 16–022
adjudication, 16–024
architect's powers and duties
  contractual claims, 14–060
concealment, 16–019
contractual liability, 16–014
fraud, 16–019
generally, 16–013
indemnity clauses, 3–070
latent damage, 16–017
mistake, 16–019
public procurement litigation, 15–031—15–032
substitution of parties, 16–022
tortious liability, 16–015
Liquidated damages
agreed sum is a penalty
  damages, 10–010
  forfeiture clauses, 10–007—10–008
  general principles, 10–005
  introduction, 10–004
  partial possession, 10–006
  sectional completion, 10–006
  set off, 10–009
defences to claims
  agreed sum is a penalty, 10–005—10–010
  determination, 10–020
  rescission, 10–020
determination, 10–020
partial possession, 10–006
penalty clauses
  damages, 10–010
  general principles, 10–005
  partial possession, 10–006
  sectional completion, 10–006
  set off, 10–009
rescission, 10–020
sectional completion, 10–006
set off, 10–009
Literal interpretation
extrinsic evidence, 3–023
Litigation
case management, 19–001
Civil Procedure Rules, 19–001
composite claims, 19–021
costs
  discretion, and, 19–080

counterclaims, 19–090
generally, 19–077
Part 20 party, 19–082
security, 19–078
written offer, 19–088
counterclaims
costs, 19–090
county court jurisdiction
generally, 19–014
delay, 19–055
disclosure
  'document', 19–049
  introduction, 19–049
  legal professional privilege, 19–050
documents, 19–048
evidence
  documents, 19–048
  witness statements, 19–043
expert witnesses
  reports, 19–045—19–046
global claims, 19–021
inspection, 19–049
interest
  judgment debts, on, 19–076
  rule of practice, 19–073
interim payments, 19–071
introduction, 19–001
legal professional privilege, 19–050
mediation, and
  consequences of failure to mediate, 19–009
  introduction, 19–008
  recoverable costs, 19–010
overriding objective, 19–001
Part 20 party, 19–082
Practice Directions, 19–001
pre-action protocol
  compliance, 19–006
  costs of compliance, 19–007
  defendant's response, 19–005
  elements, 19–005
  introduction, 19–003
  letter of claim, 19–005
  pre-action meeting, 19–005
preliminary issues, 19–052
preparation for trial
  disclosure, 19–049
  evidence, 19–048
  preliminary issues, 19–052
  settlement, 19–056
public procurement, and
  breach of regulations, 15–028
  introduction, 15–027
  limitation periods, 15–031—15–032
  obtaining information, 15–029—15–030
  remedies, 15–045
security for costs, 19–078
set-off
  exclusion of right, 19–064
settlement, 19–056

statements of case
　composite claims, 19–021
　global claims, 19–021
　stay of execution, 19–070
summary judgments
　interim payments, 19–071
　stay of execution, 19–070
TCC claims, 19–001
Technology and Construction Court
　introduction, 19–001
　practice, 19–012
witness statements, 19–043
Litigation (public procurement)
breach of regulations, 15–028
declarations of ineffectiveness
　excepted circumstances, 15–038
introduction, 15–027
limitation periods, 15–031—15–032
obtaining information
　before claim, 15–029
　during claim, 15–030
remedies for breach of regulations
　automatic suspension, 15–036
　Commission, and, 15–043
　common law, at, 15–045
　judicial review, 15–042
Local authorities
public procurement
　contracting authorities, 15–009
　generally, 15–046
Loss
Standard Building Contract with Quantities
　nature of provisions, 20–273
Loss of goodwill
generally, 9–013
Loss of profit
employer's breach, 9–033
generally, 9–013
Lump sum contracts
entire contracts
　generally, 4–003
　meaning, 4–003
　recovery of money paid, 4–007
　retention money, 4–005
extra work–exactly defined
　bills of quantities contract, 4–027
extra work–widely defined
　indispensably necessary works, 4–025
payment clauses
　entire contracts, 4–003
　substantial performance, 4–008
Maintenance
assignment of benefit by employers, 13–020
Measure of damages
duty to build dwelling properly, 16–007
Measurement and value contracts
extra work
　generally, 4–030

Mediation
See also **Arbitration**
consequences of failure to mediate, 19–009
generally, 17–001
introduction, 19–008
recoverable costs, 19–010
Misconduct
architects
　bribes, 14–026
　secret commissions, 14–026
Misrepresentation
collateral warranties
　introduction, 6–027
　personal contracts, 6–031
damages
　discretionary cause of action, 6–020
　introduction, 6–014
　knowledge of servants or agents, 6–019
exclusion of liability
　fair and reasonable, 6–025
　generally, 6–024
formation of contract, 2–001
fraudulent misrepresentation
　remedies, 6–012
innocent misrepresentation
　damages, 6–020
　generally, 6–007
　rescission, 6–007
introduction, 6–003
Misrepresentation Act 1967
　introduction, 6–014
rescission, 6–007
statutory provisions, 6–014
Mistake
architect's certificates, 5–039
limitation periods, 16–019
rectification
　common mistake, 12–012
　unilateral mistake, 12–013
Mixed contracts
public procurement, 15–008
Mutual mistake
formation of contract, 2–001
Natural justice
adjudication, 18–050
Negligence
exclusion clauses
　introduction, 3–065
indemnity clauses
　introduction, 3–065
limitation periods, 16–013
negligent instructions, 7–009
negligent misstatement
　generally, 7–027
physical damage
　generally, 7–004
　negligent instructions, 7–009
professional negligence
　negligent misstatement, 7–027
risk clauses, 3–065

Negligent misstatement
generally, 7–027
limitation periods
contract, 16–014
torts, 16–015
Negotiated procedure
generally, 15–013
short-form, 15–014
Non-performance
death
personal contracts, 6–031
default by contractor
delay, 6–077
default of other party
acceptance, 6–070
arbitration agreements, 6–069
contractual determination clauses, 6–072
fundamental breach, 6–066
generally, 6–060
frustration
recovery of sums, 6–047
illegality
generally, 6–048
return of goods, 6–051
inaccurate statements
misrepresentation, 6–003
misrepresentation
introduction, 6–003
statutory provisions, 6–014
repudiation
acceptance, 6–070
arbitration agreements, 6–069
contractual determination clauses, 6–072
fundamental breach, 6–066
generally, 6–060—
Notices
time of the essence, 8–009
Nuisance
liability to third parties
pipelines, 11–046
trees, 11–044
Offers
acceptance
'battle of forms', 2–017
certainty of terms, 2–021
conduct, by, 2–027
contract to negotiate, 2–024
lengthy negotiations, 2–018
post, telex, fax or -mail, by, 2–028
price, 2–023
'subject to contract', 2–020
count-offer, 2–012
letter of intent, 2–008
rejection, 2–011
tenders
costs, 2–004
generally, 2–003
Omissions
extra work, 4–033

On-demand bonds
generally, 11–036
Open procedure
public procurement, 15–013
Ordinary meaning
interpretation of contracts, 3–022
Overriding objective
litigation, 19–001
"Part B services contracts"
public procurement, 15–012
Partial possession
liquidated damages, 10–006
Party walls
dispute resolution
appeals, 16–063
introduction, 16–059
remedies, 16–065
"Pay-when-paid"
introduction, 3–040
Payments
architect's certificates, without
prevention by employer, 5–020
bills of quantities, 4–027
conditional payments
introduction, 3–040
date of payment
introduction, 3–040
entire contracts
generally, 4–003
meaning, 4–003
recovery of money paid, 4–007
retention money, 4–005
extra work
bills of quantities contract, 4–027
indispensably necessary works, 4–025
lump sum contracts–exactly defined,
4–027
lump sum contracts–widely defined,
4–025
measurement and value contracts, 4–030
omissions, 4–033
rate, 4–053
implied terms
date of payment, 3–040
introduction, 3–040
notice of amount, 3–040
prohibiting conditional payments, 3–040
stage payments, 3–040
suspension for non-payment, 3–040
withholding payment, 3–040
indispensably necessary works, 4–025
interim certificates, 20–234
interpretation, 3–065
loss and expense, 20–273
lump sum contracts
entire contracts, 4–003
substantial performance, 4–008
measurement and value contracts
extra work, 4–030

non-completion
  acceptance of work done, 4–012
  unpaid instalments, 4–014
notice of amount
  introduction, 3–040
notice of intention to withhold payment
  introduction, 3–040
quantum meruit
  assessment of reasonable sum, 4–021
  generally, 4–020
  meaning, 4–020
  reasonable sum, 4–021
recovery of money paid, 4–007
retention money
  generally, 4–005
stage payments
  introduction, 3–040
  Standard Building Contract with Quantities
    interim certificates, 20–234
    loss and expense, 20–273
substantial performance, 4–008
suspension for non-payment
  introduction, 3–040
withholding payment
  generally, 18–060
  introduction, 3–040
Penalty clauses
damages, 10–010
general principles, 10–005
partial possession, 10–006
sectional completion, 10–006
set off, 10–009
Peremptory orders
arbitration, 17–040
Performance
adjustment of completion, 20–100
liquidated damages, 20–120
possession, 20–056
practical completion, 20–120
Personal injury
Standard Building Contract with Quantities
  contractors' liability, 20–313
Pipelines
nuisance, 11–046
Plant and machinery
See also **Equipment**
disruption, and, 8–063
Possession
Standard Building Contract with Quantities, 20–056
PPC 2000 contract
generally, 1–024
Practical completion
Standard Building Contract with Quantities, 20–120
Practice Directions
litigation, 19–001

Pre-action protocol
compliance, 19–006
costs of compliance, 19–007
defendant's response, 19–005
elements, 19–005
introduction, 19–003
letter of claim, 19–005
pre-action meeting, 19–005
Preliminary issues
litigation, 19–052
Pricing
generally, 2–023
Professional conduct
architects, 14–008
Professional negligence
negligent misstatement, 7–027
Promissory estoppel
generally, 12–004
Public procurement
abnormally low tenders, 15–018
award criteria, 15–015—15–016
award decision, 15–020
award procedures, 15–013—15–014
breach of regulations
  generally, 15–028
  limitation periods, 15–031—15–032
  obtaining information, 15–029—15–030
  remedies, 15–045
Commission remedies, 15–043
Community legislation
  Directives, 15–002
  introduction, 15–001
  Public Contracts Regulations, 15–011
  regulations, 15–002
  TFEU, 15–001
competitive dialogue procedure, 15–014
contracting authorities, 15–009
declarations of ineffectiveness
  excepted circumstances, 15–038
Directives, 15–002
dynamic purchasing systems, 15–025
electronic auctions, 15–025
electronic procurement, 15–025
equal treatment, 15–028
excluded contracts, 15–011
implied contracts, 15–045
ineffectiveness declarations
  excepted circumstances, 15–038
interim orders, 15–036
judicial review, 15–042
late tenders, 15–019
limitation periods, 15–031—15–032
litigation
  breach of regulations, 15–028
  introduction, 15–027
  limitation periods, 15–031—15–032
  obtaining information, 15–029—15–030
  remedies, 15–045
local government level, at, 15–046
material variations, 15–022

mixed contracts, 15–008
negotiated procedure
  generally, 15–013
  short-form, 15–014
obtaining information
  before claim, 15–029
  during claim, 15–030
open procedure, 15–013
Part B services contracts, 15–012
public contracts
  abnormally low tenders, 15–018
  award criteria, 15–015—15–016
  award decision, 15–020
  award procedures, 15–013—15–014
  background, 15–002
  breach of regulations, 15–028
  concession contracts, 15–012
  contracting authorities, 15–009
  electronic procurement, 15–025
  excluded contracts, 15–011
  late tenders, 15–019
  lesser regime, 15–012
  litigation, 15–027
  material variations, 15–022
  mixed contracts, 15–008
  services contracts, 15–007
  supply contracts, 15–006
  thresholds, 15–010
  utilities, 15–026
  works contracts, 15–005
public works concession contracts, 15–012
public works contracts, 15–005
Regulations, 15–002
remedies for breach of regulations
  automatic suspension, 15–036
  Commission, and, 15–043
  common law, at, 15–045
  judicial review, 15–042
restricted procedure, 15–013
subsidised housing scheme works contracts, 15–012
suspension, 15–036
thresholds, 15–010
Treaty on the Functioning of the European Union, 15–001
utilities, 15–026
'works', 15–005
Public service contracts
  abnormally low tenders, 15–018
  award criteria, 15–015—15–016
  award decision, 15–020
  award procedures, 15–013—15–014
  background, 15–002
  breach of regulations, 15–028
  contracting authorities, 15–009
  electronic procurement, 15–025
  excluded contracts, 15–011
  introduction, 15–007
  late tenders, 15–019
  lesser regime, 15–012

litigation, 15–027
material variations, 15–022
thresholds, 15–010
utilities, 15–026
Public supply contracts
  abnormally low tenders, 15–018
  award criteria, 15–015—15–016
  award decision, 15–020
  award procedures, 15–013—15–014
  background, 15–002
  breach of regulations, 15–028
  contracting authorities, 15–009
  electronic procurement, 15–025
  excluded contracts, 15–011
  introduction, 15–006
  late tenders, 15–019
  lesser regime, 15–012
  litigation, 15–027
  material variations, 15–022
  thresholds, 15–010
  utilities, 15–026
Public works concession contracts
  public procurement, 15–012
Public works contracts
  abnormally low tenders, 15–018
  award criteria, 15–015—15–016
  award decision, 15–020
  award procedures, 15–013—15–014
  background, 15–002
  breach of regulations, 15–028
  concession contracts, 15–012
  contracting authorities, 15–009
  electronic procurement, 15–025
  excluded contracts, 15–011
  introduction, 15–005
  late tenders, 15–019
  lesser regime, 15–012
  litigation, 15–027
  material variations, 15–022
  thresholds, 15–010
  utilities, 15–026
Quality
  implied terms, 3–052
Quantum meruit
  assessment of reasonable sum, 4–021
  generally, 4–020
  innocent misrepresentation, 6–007
  meaning, 4–020
  reasonable sum, 4–021
  types of claim, 4–020
Reasonableness
  extrinsic evidence, 3–023
Rectification
  arbitration, 17–013
  common mistake, 12–012
  expert determination, 12–015A
  unilateral mistake, 12–013
Relief
  forfeiture clauses, 11–011

Index

Repudiation
acceptance, 6–070
arbitration agreements, 6–069
contractor
  delay, 6–077
contractual determination clauses, 6–072
fundamental breach, 6–066
generally, 6–060
Rescission
generally, 12–009
innocent misrepresentation, 6–007
liquidated damages, 10–020
Restitutio in integrum
innocent misrepresentation, 6–007
Restricted procedure
public procurement, 15–013
Retention money
architect's certificates, 5–013
entire contracts, 4–005
Retention of title
construction materials and equipment, 11–014
RIBA conditions of engagement
generally, 1–024
indemnity clause, 3–075
Risk clauses
loss caused by negligence, and, 3–065
Satisfactory quality
implied terms, 3–052
Scheme for Construction Contracts
introduction, 3–040
Secret commission
architect, 14–026
Security for costs
litigation, 19–078
Seizure
insolvency of contractors, 16–041
Serious irregularity
arbitration, 17–057
Set-off
exclusion of right, 19–064
limitation periods, 16–013
subcontractors, 13–058
Setting aside
arbitration awards, 17–057
Settlement
litigation, 19–056
Sewers and drains
building regulations, 16–025
Specific performance
impossibility, 12–017A
introduction, 12–017
Standard Building Contract with Quantities
adjudication, 20–405
adjustments
  completion, 20–100
carrying out the works
  adjustment of completion, 20–100
  liquidated damages, 20–120
  possession, 20–056

practical completion, 20–120
certificates
  interim certificates, 20–234
damage to property
  contractors' liability, 20–313
definitions
  commentary, 20–029
dispute resolution
  adjudication, 20–405
  introduction, 20–019
final certificates
  conclusive effect, 20–036
injury
  contractors' liability, 20–313
insolvency of contractor, 20–374
insurance
  contractors' liability, 20–313
interim certificates
  payment, 20–234
loss and expense
  nature of provisions, 20–273
overview
  design and build contracts, 1–028
  generally, 1–022—1–024
payments
  interim certificates, 20–234
  loss and expense, 20–273
performance
  adjustment of completion, 20–100
  liquidated damages, 20–120
  possession, 20–056
  practical completion, 20–120
personal injury
  contractors' liability, 20–313
possession, 20–056
practical completion, 20–120
property damage
  contractors' liability, 20–313
termination
  general provisions, 20–362
  insolvency of contractor, 20–374
Standard forms of contract
ACE conditions of engagement, 1–024
Engineering and Construction Contract
  introduction, 1–022
FIDIC conditions of contract
  introduction, 1–024
GC/Works, 1–023
generally, 1–022—1–024
I Chem E, 1–023
ICE conditions of contract
  generally, 21–001
  introduction, 1–022
MF/1, 1–023
NEC 3
  introduction, 1–022
PPC 2000 contract, 1–024
RIBA conditions of engagement, 1–024
Statements of case
composite claims, 19–021

Index

global claims, 19–021
Stay of execution
  adjudication, 18–034
  litigation, 19–070
Stay of proceedings
  arbitration
    no dispute, 17–031
    valid arbitration agreement, 17–030
Subcontractors
  access to site, 13–057
  damages, 13–060
  relationship with main contractor
    access to site, 13–057
    damages, 13–060
    introduction, 13–054
    set-off, 13–058
  set-off, 13–058
  site access, 13–057
Subject to contract
  formation of contract, 2–020
"Subsidised housing scheme works contracts"
  public procurement, 15–012
Substantial performance
  generally, 4–008
Substitution of parties
  limitation periods, 16–022
Summary judgments
  adjudication, 18–033
  interim payments, 19–071
  stay of execution, 19–070
Sureties
  formalities of contract, 2–035
  liability, 11–026
Suspension
  non-payment
    introduction, 3–040
Tax
  damages for breach of contract, 9–020
TCC claims
  generally, 19–001
Technology and Construction Court
  introduction, 19–001
  practice, 19–012
Tenders
  costs, 2–004
  generally, 2–003
Termination
  Standard Building Contract with Quantities
    general provisions, 20–362
    insolvency of contractor, 20–374

Third parties
  exclusion clauses, 3–073
Third party losses
  damages for breach of contract, 9–017
Third party rights
  arbitration, 17–012
  exclusion clauses, 3–073
Thresholds
  public procurement, 15–010
Time of the essence
  notices, 8–009
Tortious liability
  limitation periods, 16–015
Trees
  nuisance, 11–044
Unfair contract terms
  exclusion clauses
    contractual liability, 3–078
    introduction, 3–075
Unilateral mistake
  rectification, 12–013
Unincorporated associations
  capacity, 2–047
Variation
  contract works, of
    bills of quantities contract, 4–027
    indispensably necessary works, 4–025
    lump sum contracts–exactly defined, 4–027
    lump sum contracts–widely defined, 4–025
    measurement and value contracts, 4–030
    omissions, 4–033
    rate, 4–053
  generally, 12–009
Vesting clauses
  construction materials and equipment
    generally, 11–015
Vicarious performance
  assignment of burden by contractors, 13–004
Waiver
  generally, 12–005
Warranties
  assignment of benefit by employers, 13–021
Wasted expenditure
  employer's breach, 9–038
  generally, 9–014
Witness statements
  litigation, 19–043